Bio Engineering and Self-Assembly

Bio Engineering and Self-Assembly

P. R. Gilbert

ANMOL PUBLICATIONS PVT. LTD.
NEW DELHI - 110 002 (INDIA)

ANMOL PUBLICATIONS PVT. LTD.
H.O.: 4374/4B, Ansari Road, Darya Ganj
New Delhi-110 002 (India)
Ph.: 23278000, 23261597
B.O.: 1015, Ist Main Road, BSK IIIrd Stage
IIIrd Phase, IIIrd Block
Bangalore - 560 085 (Karnataka)
Ph: 080-41723429, Tel/Fax: 080-2672 3604
Visit us at: www.anmolpublications.com

Bio Engineering and Self-Assembly

First Published, 2008

ISBN 978-81-261-3276-8

PRINTED IN INDIA

Printed at Mehra Offset Press, Delhi.

Contents

Preface

This book entitled "Biotech Engineering and Self-Assembly" is designed to acquaint readers with the modern biotech engineering, nano-biotechnology and self-assembly at the global level. Bioengineering is the branch of engineering science in which biological science is used to study the relation between workers and their environments. It deals with the application of engineering principles to the fields of biology and medicine, as in the development of aids or replacements for defective or missing body organs and is also called biomedical engineering. Protein engineering is the application of science, mathematics, and economics to the process of developing useful or valuable proteins. It is a young discipline, with much research currently taking place into the understanding of protein folding and protein recognition for protein design principles. There are two general strategies for protein engineering. The first is known as *rational design*, in which the scientist uses detailed knowledge of the structure and function of the protein to make desired changes. This has the advantage of being generally inexpensive and easy, since site-directed mutagenesis techniques are well-developed. However, there is a major drawback in that detailed structural knowledge of a protein is often unavailable, and even when it is available, it can be extremely difficult to predict the effects of various mutations. Computational protein design algorithms seek to identify amino acid sequences that have low energies for target structures. While the sequence-conformation space that needs to be searched is large, the most challenging requirement for computational protein design is a fast, yet accurate, energy function that can distinguish optimal sequences from similar suboptimal ones. Using computational methods, a protein with a novel fold has been designed, as well as sensors for un-natural molecules. The second strategy is known as directed evolution. This is where random mutagenesis is applied to a protein, and a selection regime is used to pick out variants that have the desired qualities. Further rounds of mutation and selection are then applied. This method mimics natural evolution and generally produces superior results to rational design.

The great advantage of directed evolution techniques is that they require no prior structural knowledge of a protein, nor it is necessary to be able to predict what effect a given mutation will have. Indeed, the results of directed evolution experiments are often surprising in that desired changes are often caused by mutations that no one would have expected. The drawback is that they require high-throughput, which is not feasible for all proteins. Large amounts of recombinant DNA must be mutated and the products screened for desired qualities. The sheer number of variants often requires expensive robotic equipment to automate the process. Furthermore, not all desired activities can be easily screened for. Rational design and directed evolution techniques are not mutally exclusive; good researchers will often apply both. In the future, more detailed knowledge of protein structure and function, as well as advancements in high-throughput technology, will greatly expand the capabilities of protein engineering. Very basically, food-related biotechnology is the process by which a specific gene or group of genes with desirable traits are removed from the DNA of one plant or animal cell and spliced into that of another. Such beneficial genes might come from animals, (friendly) bacteria, fish, insects, plants and even humans. In some instances, genes that create problems (such as the natural softening of a tomato) are simply removed and not replaced. Tomatoes, for example, are generally picked green and gas-ripened later because, during shipping, they would become soft, bruised and unmarketable. A bioengineered tomato, however, can be picked ripe and shipped without softening. The objective of food biotechnology is to develop insect- and disease-resistant, shipping- and shelf-stable foods with improved appearance, texture and flavor. Additionally, biotechnology advocates say that the process will produce plants that are resistant to adverse weather conditions such as drought and frost, thereby increasing food production in previously prohibitive climate and soil conditions. They also envision increasing nutrient levels and decreasing pesticide usage through biotechnology. On the other hand, critics argue that, because biotechnology is producing new foods not previously consumed by humans, the changes and potential risks relating to such things as toxins, allergens and reduced nutrients are unpredictable. They also worry that, because genetically altered foods are not required to be labeled, people with religious or lifestyle dietary restrictions might unintentionally consume prohibited foods.

This book consists of—Biotech Engineering: An Introductory Overview; Biotechnology and Nano World: The Relationship; DNA Microarrays Technology and Its Uses; and Self-Assembly and Nano-Biotechnology: Crucial Issues etc. besides a large body of list of acronyms, glossary of relevant terms, extensive bibliography, list of websites and links, for further referencing and research.

—Editor

List of Acronyms

AAIP	—	American Association of Investigative Pathologists
ABI	—	Applied Biosystems, Inc.
ABRF	—	Association of Biomolecular Resource Facilities
ACMG	—	American College of Medical Genet.
ACS	—	American Chemical Society
ADA	—	Americans with Disabilities Act
AES	—	American Electrophoresis Society
AFMR	—	American Federation of Medical Research
AGSG	—	Alliance of Genetic Support Groups (now Genetic Alliance)
AGT	—	Association of Genetic Technologists
BER	—	Biological and Environmental Research
BIO	—	Biotechnology Industry Organization
BNL	—	Brookhaven National Laboratory
BSCS	—	Biological Sciences Curriculum Study
BS/SCF	—	Biological Sequence/Structure Computational Facility

BTCI	—	BioPharmaceutical Technology Center Institute
CATCMB	—	Center for Advanced Training in Cell and Molecular Biology
CCM	—	Chromosome Coordinating Meeting
CLMA	—	Clinical Laboratory Management Association
CMT	—	Charcot-Marie-Tooth
CONTIG	—	Consortium of Teachers in Genetics
CORN	—	Council of Regional Networks for Genetic Services
CRADA	—	Cooperative Research and Development Agreement
DM	—	Myotonic Dystrophy
DMD	—	Duchenne Muscular Dystrophy
DOE	—	Department of Energy
EEOC	—	Equal Employment Opportunity Commission
EL.B.A.	—	ELectronics and Biotechnology Advanced
EMBL	—	European Mol. Biol. Lab.
EMBO	—	European Molecular Biology Organisation
EMG	—	Encyclopedia of the Mouse Genome
EMS	—	Environmental Mutagen Society
EORTC	—	European Organization for Research and Treatment of Cancer
ERDA	—	Energy Research and Development Administration
ERI	—	Eleanor Roosevelt Institute

ES	—	Embryonic Stem
ESF	—	European Science Foundation
GMCRF	—	General Motors Cancer Res. Foundation
GMD	—	Genomic Map Design
GPI	—	Genetics and Public Issues Program
GRAIL	—	Gene Recognition and Analysis Internet Link
GRC	—	Gordon Res. Conf.
HERAC	—	Health and Environmental Research Advisory Committee
HGCC	—	Human Genome Coordinating Committee
HGM	—	Human Genome Meeting
HGMIS	—	Human Genome Management Information System
HGP	—	Human Genome Project
HHMI	—	Howard Hughes Medical Institute
HICSS	—	Hawaii Intl. Conf. on Systems Sci.
HLA	—	Human Leukocyte Antigen
HMDP	—	Homology Database
HNPCC	—	Hereditary Nonpolyposis Colorectal Cancer
IBC	—	International Business Communications
IBEX	—	International Biotechnology EXpo
IBF	—	International Business Forum
IBI	—	Institute for Biotechnology Information

ICES	—	International Council of Electrophoresis Societies
ICGEB	—	International Centre for Genetic Engineering and Biotechnology
ICHG	—	International Congress of Human Genetics
ICPEMC	—	International Commission on Protection Against Environmental Mutagens and Carcinogens
ICRF	—	Imperial Cancer Research Fund
IEEE	—	Institute of Electrical and Electronics Engineers
IG	—	IntelliGenetics
IGES	—	International Genetics Epidemiology Societies
IJCAI	—	International Joint Conference on Artificial Intelligence
IMA	—	Institute for Mathematics and its Applications
IMACS	—	International Association for Mathematics and Computers In Simulation
IMAGE	—	Integrated Molecular Analysis of Gene Expression
IMGS	—	Intl. Mammalian Genome Society
INRIA	—	French Natl. Inst. for Research in Computer Science and Control
IOM	—	Institute of Medicine
IOR	—	Institute of Religion
ISAG	—	Intl. Society for Animal Genetics

ISMB	—	Intelligent Systems for Molecular Biology
ISONG	—	Intl. Soc. of Nurses in Genet.
ISQL	—	Interactive Standard Query Language
ISTR	—	Institute for Science Training and Research
IUBMB	—	Intl. Union of Biochemistry and Molecular Biology
IU	—	Indiana University
LBNL	—	Lawrence Berkeley National Laboratory
LLNL	—	Lawrence Livermore National Laboratory
LTI	—	Life Technologies, Inc.
MBC	—	Massachusetts Biotechnology Council
MBL	—	Marine Biological Laboratory
MCD	—	Mouse Cytogenetic Database
MDA	—	Muscular Dystrophy Assoc.
MEMS	—	Microelectromechanical Systems
MGC	—	Mouse Genome Conference
MGD	—	Mouse Genome Database
MGI	—	Microbial Genome Initiative
MHC	—	Major Histocompatibility Complex
MIMBD	—	Meet. on the Interconnection of Mol. Biol. Databases
MIT	—	Massachusetts Institute of Technology

MMRF	—	Marshfield Medical Research Foundation
MNBWS	—	Miami Nature Biotechnology Winter Symposium
MOD	—	March of Dimes
MOU	—	Memorandum of Understanding
MRC	—	Medical Research Council
NAPBC	—	Natl. Action Plan on Breast Cancer
NAS	—	Natl. Academy of Sciences
NASA	—	Natl. Aeronautics and Space Administration
NBAC	—	Natl. Bioethics Advisory Commission
NCGR	—	National Center for Genome Resources
NCI	—	Natl. Cancer Institute
NCSA	—	National Center for Supercomputing Applications
NCSL	—	National Conference of State Legislatures
NCSU	—	North Carolina State University
NEH	—	National Endowment for the Humanities
NFCR	—	National Foundation for Cancer Research
NFID	—	National Foundation for Infectious Diseases
NHGRI	—	Natl. Human Genome Res. Inst.
NICHD	—	National Institute of Child Health and Human Development

NIGMS	—	National Institute of General Medical Sciences
NIH	—	National Institutes of Health
NIGMS	—	Natl. Institute of General Medical Sciences
NIST	—	Natl. Inst. of Standards and Technology
NLGLP	—	National Laboratory Gene Library Project
NLM	—	National Library of Medicine
NMHCC	—	National Managed Health Care Congress
NORD	—	National Organization for Rare Disorders
NRC	—	National Research Council
NRSA	—	National Research Service Award
NSF	—	Natl. Sci. Foundation
NSGC	—	National Society of Genetic Counselors
OHER	—	Office of Health and Environmental Research
OMIM	—	Online *Mendelian Inheritance in Man*
OPRR	—	Office of Protection from Res. Risks
ORAU	—	Oak Ridge Associated Universities
ORISE	—	Oak Ridge Inst. for Science and Education
ORF	—	open reading frame
ORNL	—	Oak Ridge National Laboratory
OSU	—	Oregon State University

\SCE	—	School of Continuing Education
SCI	—	Society of Chemical Industry
SCW	—	Single-chromosome Workshop
SDC	—	San Diego Conference
SNL	—	Sandia National Laboratory
SNP	—	Single Nucleotide Polymorphism
SQL	—	Standard Query Language
SSR	—	Simple Sequence Repeat
STM	—	Scanning Tunneling Microscope
TIGR/NIST	—	The Inst. for Genomic Res./Natl. Inst. of Standards and Technol.
UCB	—	University of California, Berkeley

List of Glossary of Terms

Aromatic: A term used to describe cyclic pi-bonded structures of special stability.

Asexual reproduction: Nonsexual means of reproduction which can include grafting and budding.

Autoproductivity: The ability of a system, under external control, to automatically produce an identical copy of itself.

Bacteriophage (phage or phage particle): A virus that infects bacteria. Altered forms are used as vectors for cloning DNA.

Biochauvinism: The prejudice that biological systems have an intrinsic superiority that will always give them a monopoly on self-reproduction and intelligence.

Biocompatible coated materials: Biocompatible materials usually used in dental and bone implants that enhance biologic fixation, thereby increasing the bond strength between the coated material and bone, and minimize possible biological effects that may result from the implant itself. MeSH, 1999

Biocompatible materials: Synthetic or natural materials, other than drugs, that are used to replace or repair any body tissue or bodily function. MeSH, 1973

Biomimetic materials: Materials fabricated by BIOMIMETICS techniques, i.e., based on natural processes found in biological systems. [MeSH 2003] (2) Materials that imitate, copy, or learn from nature.

Biomolecular Nanotechnology: Nanotechnology existing in living systems and resulting from our ability to use biomolecules as components for molecular nanotechnology.

Biomotors: Driven by energy sources such as *adenosine triphosphate (ATP)* for chemical transduction and other processes. These biomotors are considered to be biomolecular and are discussed in the body of this report, but strictly speaking they do not conform to the panel's definition of self- assembly.

Blot, western: A pattern of proteins transferred to a nitrocellulose membrane from a gel. The gel has undergone electrophoresis to separate fragments according to size. The proteins are arranged in different lanes for each sample. Each lane contains bands which are fragments of different sizes. Labeled antibody probes are often used to visualize particular bands.

Bone substitutes: Synthetic or natural materials for the replacement of bones or bone tissue. They include hard tissue replacement polymers, natural coral, hydroxyapatite, beta- tricalcium phosphate, and various other biomaterials. The bone substitutes as inert materials can be incorporated into surrounding tissue or gradually replaced by original tissue.

Brownian Assembly: Brownian motion in a fluid brings molecules together in various position and orientations. If molecules have suitable complementary surfaces, they can bind, assembling to form a specific structure. Brownian assembly is a less paradoxical name for self-assembly (how can a structure assemble itself, or do anything, when it does not yet exist?).

Bt (*Bacillus thuringiensis*): A soil bacterium that produces insecticidal proteins. There are several different kinds of proteins produced by different strains of Bt. Some are effective against larvae of moths and butterflies. Others are effective against larvae of beetles. The Bt protein has been introduced into various crops as a built-in insecticide.

Catalyst: A substance that promotes a chemical reaction by lowering the activation energy of a chemical reaction, but which itself remains unaltered at the end of the reaction. (See Catalytic antibody, Catalytic RNA.)

Catalytic antibody (abzyme): An antibody selected for its ability to catalyze a chemical reaction by binding to and stabilizing the transition state intermediate.

Cell pharmacology: Delivery of drugs by medical nanomachines to exact locations in the body.

Chromosome: A single DNA molecule, a tightly coiled strant of DNA, condensed into a compact structure in vivo by complexing with accessory histones or histone-like proteins. Chromosomes exist in pairs in higher eukaryotes. (See Chromosome walking.) A linear structure in the nucleus of plants and animals that is visible in light microscopy when stained. The chromosome is a single, long, linear molecule of DNA and associated proteins. Bacteria have a single circular chromosome; other organisms may have many linear chromosomes.

Cistron: A DNA sequence that codes for a specific polypeptide; a gene. See DNA, Gene.

Clone: An exact genetic replica of a specific gene or an entire organism.

Cloning vector: A DNA molecule capable of autonomous

replication within the cloning host cell (e.g *E. coli*). The vectors contain restriction enzyme sites for insertion of foreign DNA. Cloning vectors are derived from bacterial plasmids, bacteriophages, or viruses.

Cloning: The mitotic division of a progenitor cell to give rise to a population of identical daughter cells or clones. (See Directional cloning, Megabase cloning, Molecular cloning, Subcloning.)

Coat protein (capsid): The coating of a protein that enclosed the nucleic acid core of a virus.

Codon: A group of three nucleotides that specifies addition of one of the 20 amino acids during translation of an mRNA into a polypeptide. Strings of codons form genes and strings of genes form chromosomes. Set of three nucleotides that specify a particular amino acid during protein synthesis.

Coenzyme (cofactor): An organic molecule, such as a vitamin, that binds to an enzyme and is required for its catalytic activity.

Colony: A group of identical cells (clones) derived from a single progenitor cell.

Combinatorial materials design: Uses computing power (sometimes together with massive parallel experimentation) to screen many different materials possibilities to optimize properties for specific applications (e.g., catalysts, drugs, optical materials).

Commensalism: The close association of two or more dissimilar organisms where the association is advantageous to one and doesn't affect the other(s). See Parasitism, Symbiosis.

Competency: An ephemeral state, induced by treatment with cold cations, during which bacterial cells are capable of uptaking foreign DNA.

Complementary DNA or RNA: The matching strand of a DNA or RNA molecule to which its bases pair. (See DNA, RNA.)

Complementary nucleotides: Members of the pairs adenine-thymine, adenine-uracil, and guaninecytosine that have the ability to hydrogen bond to one another. (See nucleotide.)

Concatemer: A DNA segment composed of repeated sequences linked end to end.

Configuration space: A mathematical space describing the three-dimensional configuration of a system of particles (e.g., atoms in a nanomechanical structure) as a single point; the configuration space for an *N* particle system has 3*N* dimensions.

Conformation: The shape of a molecule. A molecular geometry that differs from other geometries chiefly by rotation about single or triple bonds; distinct conformations (termed con formers) are associated with distinct potential wells. Typical biomolecules and products of organic synthesis can interconvert among many conformations; typical diamondoid structures are locked into a single potential well, and thus lack conformational flexibility.

Crystal Lattice: The regular three-dimensional pattern of atoms in a crystal.

Crystal structure: For crystalline materials, the manner in which atoms or ions are arrayed in space. It is defined in terms of the unit cell geometry and the atom positions within the unit cell.

Crystallescence: In medical nanorobotics, the crystallization of solid solute that is offloaded by nanorobot sorting rotors at a concentration that exceeds the solvation capacity of the surrounding solvent.

Culture: A particular kind of organism growing in a laboratory medium.

Curie temperature (also Curie point) (Tc): The temperature above which a ferromagnetic or ferrimagnetic material becomes paramagnetic. For iron the Curie point is 760oC and for nickel 356oC.

DNA, repetitive: Fragments of DNA that appear in multiple copies in a single individual.

DNA: Deoxyribonucleic acid, the molecule that stores genetic information. Composed of two complementary strands. See base pairs.

Dominant gene: A gene whose phenotype is when it is present in a single copy.

Dominant(-acting) oncogene: A gene that stimulates cell proliferation and contributes to oncogenesis when present in a single copy. (See Oncogene.)

Double helix: Describes the coiling of the antiparallel strands of the DNA molecule, resembling a spiral staircase in which the paired bases form the steps and the sugar-phosphate backbones form the rails.

Encapsidation: Process by which a virus' nucleic acid is enclosed in a capsid. See Coat protein.

Endocytosis: Uptake of material into a cell by an invagination of the plasma membrane and its internalization in a membrane-bounded vesicle.

Endophyte: An organism that lives inside another.

Endoplasmic reticulum (ER): Labyrinthine, membrane-bounded compartment in the cytoplasm of eucaryotic cells, where lipids are synthesized and membrane-bound proteins are made.

Endosome: Membrane-bounded organelle in animal cells that carries materials newly ingested by endocytosis and passes many of them on to lysosomes for degradation.

GEM: A genetically engineered microorganism.

Gene amplification: The presence of multiple genes. Amplification is one mechanism through which proto-oncogenes are activated in malignant cells.

Gene frequency: The percentage of a given allele in a population of organisms. See Allele.

Gene gun: Transformation technique that uses accelerated particles coated with DNA to introduce foreign DNA into recipient plant.

Gene insertion: The addition of one or more copies of a normal gene into a defective chromosome.

Gene linkage: The hereditary association of genes located on the same chromosome.

Gene modification: The chemical repair of a gene's defective DNA sequence. See DNA.

Gene pool: The totality of all alleles of all genes of all individuals in a particular population.

Gene: The basic unit of inheritance. A segment of DNA that codes for a particular protein. A locus on a chromosome that encodes a specific protein or several related proteins. It is considered the functional unit of heredity. (See Dominant gene, Fusion gene, Gene amplification, Gene expression, Gene flow, Gene pool, Gene splicing, Gene translocation, Recessive gene, Regulatory gene.)

Genetic assimilation: Eventual extinction of a natural species as massive pollen flow occurs from another related species and the older crop becomes more like the new crop. See Gene flow.

Genetic code: The three-letter code that translates nucleic acid sequence into protein sequence. The relationships between the nucleotide base-pair triplets of a messenger RNA molecule and the 20 amino acids that are the building blocks of proteins. See Base pair, Nucleic acid, Nucleotide. The genetic information in DNA is encoded with four different nucleotide bases: A, C, G, and T (see base pair). A set of three consecutive nucleotide bases constitutes a codon. A codon specifies a particular amino acid that is added during synthesis of a protein.

Genetic disease: A disease that has its origin in changes to the genetic material, DNA. Usually refers to diseases that are inherited in a Mendelian fashion, although noninherited forms of cancer also result from DNA mutation.

Genetic drift: Random variation in gene frequency from one generation to another.

Genetic engineering: Genetic engineering is the process of manually adding new DNA on a molecular level with the goal of adding one or more new traits that are not already found in that organism. For the purposes of this web site, genetic engineering includes recombinant DNA and gene splicing technologies.

Genetic linkage map: A linear map of the relative positions of genes along a chromosome. Distances are established by linkage analysis, which determines the frequency at which two gene loci become separated during chromosomal recombination. (See Mapping.)

Genetic marker: A gene or group of genes used to "mark" or track the action of microbes.

Genetic modification: The selective and purposeful alteration of genes by humans. Genetic modification includes methods of incorporating new genes or traits into an organism through genetic engineering, mutagenesis and traditional breeding methods.

Genome: The genetic complement contained in the chromosomes of a given organism, usually the haploid chromosome state. The full chromosome set containing all the genes of a particular individual.

Genomic library: A library composed of fragments of genomic DNA. (See Library.)

Herbicide: Any substance that is toxic to plants; usually used to kill specific unwanted plants.

Heterochromatin: Dark-stained regions of chromosomes thought to be for the most part genetically inactive.

Heteroduplex: A double-stranded DNA molecule or DNA-RNA hybrid, where each strand is of a different origin.

Heterogeneous nuclear RNA (hnRNA): The name originally given to large RNA molecules found in the nucleus, which are now known to be unedited mRNA transcripts, or pre-mRNAs. (See RNA.)

Homologous chromosomes: Chromosomes that have the same linear arrangement of genes—a pair of matching chromosomes in a diploid organism. See Chromosomes.

Homologous recombination: The exchange of DNA fragments between two DNA molecules or chromatids of paired chromosomes (during crossing over) at the site of identical nucleotide sequences.

Homozygote: An organism whose genotype is characterized by two identical alleles of a gene. See Allele, Genotype.

Host: An organism that contains another organism.

Human Genome Project: A project coordinated by the

National Institutes of Health (NIH) and the Department of Energy (DOE) to determine the entire nucleotide sequence of the human chromosomes. (See NIH.)

Human growth hormone (HGH, somatotrophin): A protein produced in the pituitary gland that stimulates the liver to produce somatomedins, which stimulate growth of bone and muscle.

Hybridization: The hydrogen bonding of complementary DNA and/or RNA sequences to form a duplex molecule. (See Northern hybridization, Southern hybridization.)

Hybridoma: A hybrid cell, composed of a B lymphocyte fused to a tumor cell, which grows indefinitely in tissue culture and is selected for the secretion of a specific antibody of interest.

Interferon: A family of small proteins that stimulate viral resistance in cells.

Intergenic regions: DNA sequences located between genes that comprise a large percentage of the human genome with no known function.

Introgression: Backcrossing of hybrids of two plant populations to introduce new genes into a wild population.

Intron: A noncoding DNA sequence within a gene that is initially transcribed into messenger RNA but is later snipped out. See Coding, DNA, Messenger RNA, Transcription.

Ligase (DNA ligase): An enzyme that catalyzes a condensation reaction that links two DNA molecules via the formation of a phosphodiester bond between the 3' hydroxyl and 5' phosphate of adjacent nucleotides.

Light Emitting Diodes (LEDs): Light Emitting Diodes work on a completely different concept. Traditionally LEDs are created from two semiconductors. By running current in one direction across the semiconductor the LED emits light of a particular frequency (hence a particular color) depending on the physical characteristics of the semiconductor used. The semiconductor is covered with a piece of plastic that focuses the light and increases the brightness. These semiconductors are very durable, there is no filament, they donít require much power, theyíre brighter and they last a long time. By densely packing red, blue and green LEDs next to each other on a substrate one can create a display.

Lineage: A chart that traces the flow of genetic information from generation to generation.

Linkage: The frequency of coinheritance of a pair of genes and/or genetic markers, which provides a measure of their physical proximity to one another on a chromosome. Measures the physical distance between two genes. Genes that are close together are unlikely to segregate in a sexual cross. Distant genes segregate independently are are then said to be unlinked.

Linked genes/markers: Genes and/or markers that are so closely associated on the chromosome that they are coinherited in 80% or more of cases.

Linker: A short, double-stranded oligonucleotide containing a restriction endonuclease recognition site, which is ligated to the ends of a DNA fragment.

Liposome: A type of nanoparticle made of lipids, or fat molecules, surrounding a water core. Liposomes, several of which are widely used to treat infectious diseases and cancer, were the first type of nanoparticle to be used to create therapeutic agents with novel characteristics. Membrane-bound vesicles constructed in the laboratory to transport biological molecules.

Liquid Crystal Display: LCD is the predominant technology used in flat panel displays. The principle that makes the display work is this: A crystalís alignment can be altered with an electric current. If the crystal is lined up one way ñ it will allow the light waves to pass through a polarized filter, but if the electric current alters the crystalís alignment, it will guide light so that the polarized filter blocks the light. By densely packing red, blue and green light emitting crystals next to each other on a sheet (ìcalled a substrateî), one can create a full color display. The great thing about LCD is that the crystals can be packed together closely, allowing for a higher-resolution, finer-detail display. The con is that LCDs are somewhat fragile, require a lot of power and are relatively less bright. Liquid Crystal Display — display device employing light source and electrically alterable optically active thin film.

Logarithmic phase (log or exponential growth phase): The steepest slope of the growth curve—the phase of vigorous growth during which cell number doubles every 20-30 minutes. (See Growth phase.)

Lysogen: A bacterial cell whose chromosome contains integrated viral DNA.

Lysogenic: A type or phase of the virus life cycle during which the virus integrates into the host chromosome of the infected cell, often remaining essentially dormant for some period of time. See Lysogen.

Lytic: A phase of the virus life cycle during which the virus replicates within the host cell, releasing a new generation of viruses when the infected cell lyses.

Machine-phase chemistry: The chemistry of systems in which all potentially reactive moieties follow controlled trajectories (e.g., guided by molecular machines working in vacuum).

Macroscale: Larger than nanoscale; often implies a design that humans can directly interact with; too large to be built by a single assembler (one cubic micron of diamond contains 176 billion atoms).

Magnetization (M): The total magnetic moment per unit volume of material. Also, a measure of the contribution to the magnetic flux by some material within an H field. The magnitude of M is proportional to the applied field as: M = ÷m × H, with ÷m the magnetic susceptibility.

Magnetostrictive material: A material that changes dimension in the presence of a magnetic field or generates a magnetic field when mechanically deformed.

Malignant: Having the properties of cancerous growth.

Manufacturing: The ability to make products, in this case ranging from clothing, to electronics, to medical devices, to books, to building materials, and much more.

Measurand: A physical quantity, condition, or property that is to be measured.

Meat Machine: AKA Cabinet Beast. A box containing assemblers and raw material, within which is formed meat [or whatever else it was programmed to make].

Mechanochemistry: Chemistry accomplished by mechanical systems directly controlling the reactant molecules; the formation or breaking of chemical bonds under direct mechanical control. In this volume, the chemistry of processes in which mechanical systems operating with atomic-scale precision either guide, drive, or are driven by chemical transformations. In general usage, the chemistry of processes in which energy is converted from mechanical to chemical form, or vice versa.

Mechanosynthesis: Chemical synthesis controlled by mechanical systems operating with atomic-scale precision, enabling direct positional selection of reaction sites; synthetic applications of mechanochemistry. Suitable mechanical systems include AFM mechanisms, molecular manipulators, and molecular mill systems. Processes that fall outside the intended scope of this definition include reactions guided by the incorporation of reactive moieties into a shared covalent framework (i.e., conventional intramolecular reactions), or by the binding of reagents to enzymes or enzymelike catalysts. Molecular tools with chemically specific tip structures can be used, sequentially, to modify a work piece and build a wide range of molecular structures.

Megahertz (Mhz): One million cycles per second

Meiosis: The reduction division process by which haploid gametes and spores are formed, consisting of a single duplication of the genetic material followed by two mitotic divisions.

Metabolism: The biochemical processes that sustain a living cell or organism.

Metallothionein: A protective protein that binds heavy metals, such as cadmium and lead.

Micellar nanocontainers: Block copolymer micelles are water- soluble biocompatible nanocontainers with great potential for delivering hydrophobic drugs. An understanding of their cellular distribution is essential to achieving selective delivery of drugs at the subcellular level. R. Savic, L. Luo, A. Eisenberg, D. Maysinger, Micellar nanocontainers distribute to defined cytoplasmic organelles Science 300 (5619): 615- 618, Apr. 25, 2003

Microspectrophotometry: Analytical technique for studying substances present at enzyme concentrations in single cells, in situ, by measuring light absorption. Light from a tungsten strip lamp or xenon arc dispersed by a grating monochromator illuminates the optical system of a microscope. The absorbance of light is measured (in nanometers) by comparing the difference between the image of the sample and a reference image. MeSH 1990

Microspheres: Microspheres are generally defined as small spheres made of any material and sized from about 0.5 μm to 100 μm. Similar, but smaller spheres sized 10 to 500 nm are called *nanospheres.* Ideally, microspheres are completely spherical and homogeneous in size, although less perfect particles are often termed microspheres as well. Depending on the preparation method and material used, microspheres show a typical size distribution which often deviates from the mono-sized ideal.

Microstructure: In material engineering the structural features of a material such as grain boundaries, grain size and structure, subject to observation under a microscope, selective etching etc. In MEMS microstructure unfortunately is also used to designate a micromachined feature. The last decade has seen rapid developments in the fabrication, characterization and conceptual understanding of synthetic microstructures in many different material systems including silicon, III-V and II-VI semiconductors, metals, ceramics and organics. The objective of this journal [Superlattices and Microstructures] is to provide a common interdisciplinary platform for the publication of the latest research results on all such "nanostructures" with dimensions in the range of 1-100 nm; the unifying theme here being the dimensions of these artificial structures rather than the material system in which they are fabricated.

Microsystem: A microscale machine that can sense information from the environment and act accordingly. Outside the U.S., it can also refer to microelectromechanical systems (MEMS).

Molecular assembler: A general-purpose device for molecular manufacturing, able to guide chemical reactions by positioning individual molecules to atomic accuracy (e.g. mechanosynthesis) and to construct a wide range of useful and stable molecular structures according to precise specifications. Also known as an assembler, a molecular assembler is a molecular machine that can build a molecular structure from its component building blocks.

Molecular Beam Epitaxy: [MBE] Process used to make compound (multi-layer) semiconductors. Consists of depositing alternating layers of materials, layer by layer, one type after another (such as the semiconductors gallium arsenide and aluminum gallium arsenide).

Molecular biology: The study of the biochemical and molecular interactions within living cells.

Molecular cloning:: The biological amplification of a specific DNA sequence through mitotic division of a host cell into which it has been transformed or transfected.

Molecular genetics: The study of the flow and regulation of genetic information between DNA, RNA, and protein molecules.

Molecular markers: Genetic traits which are detectable on gels or blots and can be used to construct genetic maps. Molecular markers usually have no known function. RFLPs are molecular markers.

Molecular mechanics models: Many of the properties of molecular systems are determined by the molecular potential energy function. Molecular mechanics models approximate this function as a sum of 2-atom, 3-atom, and 4-atom terms, each determined by the geometries and bonds of the component atoms. The 2-atom and 3-atom terms describing bonded interactions roughly correspond to linear springs.

Molecular Wire: A molecular wire—the simplest electronic component—is a quasi-one-dimensional molecule that can transport charge carriers (electrons or holes) between its ends. [Michael D Ward]

Molecularly imprinted polymers MIPs: A new class of materials that have artificially created receptor structures. Since their discovery in 1972, MIPs have attracted considerable interest from scientists and 3engineers involved with the development of chromatographic absorbents, membranes, sensors and enzyme and receptor mimics.

Monoclonal antibodies: Immunoglobulin molecules of single- epitope specificity that are secreted by a clone of B cells.

Multi-locus probe: A probe that hybridizes to a number

of different sites in the genome of an organism. (See Probe.)

Mutagen: Any agent or process that can cause mutations. See Mutation.

Nanoscience: Nanoscience is the science of nanoscale phenomena. This is a very complicated area, since Quantum Mechanics is often needed to describe the physical properties of atoms and molecules properly. What makes it even more complicated is that nanostructured materials can still be quite a challenge to describe Quantum Mechanically, since the equations are feindishly difficult—this is where supercomputers come in to make the job approachable. The scientific discipline seeking to increase our knowledge and understanding of nanoscale phenomena, i.e. science on the scale of 0.1 nm to 100 nm.

Nanosensor: A chemical or physical sensor constructed using nanoscale components, usually microscopic or submicroscopic in size.

National Institutions of Health (NIH): A nonregulatory agency which has oversight of research activities that the agency funds.

National Science Foundation (NSF): A nonregulatory agency which has oversight of biotechnology research activities that the agency funds.

Natural selection: The differential survival and reproduction of organisms with genetic characteristics that enable them to better utilize environmental resources.

Neophile: One who welcomes the future and who enjoys change and evolution.

Nick translation: A procedure for making a DNA probe in which a DNA fragment is treated with DNase to produce single-stranded nicks, followed by incorporation of radioactive nucleotides from the nicked sites by DNA polymerase I.

Nicked circle (relaxed circle): During extraction of plasmid DNA from the bacterial cell, one strand of the DNA becomes nicked. This relaxes the torsional strain needed to maintain supercoiling, producing the familiar form of plasmid. (See Plasmid.)

Nitrocellulose: A membrane used to immobilize DNA, RNA, or protein, which can then be probed with a labeled sequence or antibody.

Nitrogen fixation: The conversion of atmospheric nitrogen to biologically usable nitrates.

Nitrogenous bases: The purines (adenine and guanine) and pyrimidines (thymine, cytosine, and uracil) that comprise DNA and RNA molecules.

Nodule: The enlargement or swelling on roots of nitrogen-fixing plants. The nodules contain symbiotic nitrogen-fixing bacteria. See Nitrogen fixation.

Nontarget organism: An organism which is affected by an interaction for which it was not the intended recipient.

Northern hybridization: (Northern blotting). A procedure in which RNA fragments are transferred from an agarose gel to a nitrocellulose filter, where the RNA is then hybridized to a radioactive probe. (See Hybridization.)

Nuclease: Any enzyme that cuts nucleic acids. See restriction enzyme. A class of enzymes that degrades DNA and/or RNA molecules by cleaving the phosphodiester bonds that link adjacent nucleotides. In deoxyribonuclease (DNase), the substrate is DNA. In endonuclease, it cleaves at internal sites in the substrate molecule. Exonuclease progressively cleaves from the end of the substrate molecule. In ribonuclease (RNase), the substrate is RNA. In the S1 nuclease, the substrate is single-stranded DNA or RNA.

Nucleic acids: The two nucleic acids, deoxyribonucleic acid (DNA) and ribonucleic acid (RNA), are made up of long chains of molecules called nucleotides. See DNA, RNA, Nucleotides.

Nuclein: The term used by Friedrich Miescher to describe the nuclear material he discovered in 1869, which today is known as DNA.

Nucleotide: A building block of DNA and RNA, consisting of a nitrogenous base, a five-carbon sugar, and a phosphate group. Together, the nucleotides form codons, which when strung together form genes, which in turn link to form chromosomes. (See Chromosome, Codon, Complementary nucleotides, Dideoxynucleotide, DNA, Gene, Oligonucleotide, RNA.)

Nucleus: The membrane-bound region of a eukaryotic cell that contains the chromosomes. A cellular organelle in plants and animals that contains the chromosomes which in turn are composed of DNA plus protein.

Occupational Safety and Health Administration (OSHA): One of the U.S. agencies responsible for regulation of biotechnology. The major law under which the agency has regulatory powers is the Occupational Safety and Health Act.

Oligonucleotide: A DNA polymer composed of only a few nucleotides. (See Nucleotide.)

Oncogene: A gene that contributes to cancer formation when mutated or inappropriately expressed. (See

Cellular oncogene, Dominant oncogene, Immortalizing oncogene, Recessive oncogene.)

Oncogenesis: The progression of cytological, genetic, and cellular changes that culminate in a malignant tumor.

Orbital: In the approximation that each electron in a molecule has a distinct, independent wave function, the spatial distribution of an electron wave function corresponds to a molecular orbital. These, in turn, can be approximated as sums of contributions from the orbitals characteristic of the isolated atoms. An electron added to a molecule—or, similarly, one excited to a higher-energy state within a molecule—would occupy a state with a different wave function from the rest; an unoccupied state of this kind corresponds to an unoccupied molecular orbital. Orbital-symmetry effects on reaction rates arise when a reaction requires overlap between two lobes of the orbitals on each of two reagents: if the algebraic signs of the wave functions in the facing lobes do not match, bond formation between those orbitals is prohibited.

Overlapping reading frames: Start codons in different reading frames generate different polypeptides from the same DNA sequence. (See Reading frame.)

Paleontology: The study of the fossil record of past geological periods and of the phylogenetic relationships between ancient and contemporary plant and animal species.

Palindromic sequence: A DNA locus whose 5'-to-3' sequence is identical on each DNA strand. The sequence is the same when one strand is read left to right and the other strand is read right to left. Recognition sites of many restriction enzymes are palindromic. See DNA.

pAMP: Ampicillin-resistant plasmid developed for this laboratory course. (See Plasmid.)

Parasitism: The closee association of two or more dissimilar organisms where the association is harmful to at least one. See Commensalism, Parasitism, Symbiosis.

Particle bombardment: Using metal particles to blast DNA into cells for the purpose of genetic engineering. A gene gun is used to propel the metal particles.

PCR: Polymerase chain reaction. An amazingly sensitive technique that allows the specific amplification of extremely small amounts of particular DNA fragments using DNA polymerase and specific primers.

Pedigree: A diagram mapping the genetic history of a particular family.

Percent Accepted Mutation: A unit introduced by Dayhoff et al. to quantify the amount of evolutionary change in a protein sequence. 1.0 PAM unit, is the amount of evolution which will change, on average, 1% of amino acids in a protein sequence. A PAM(x) substitution matrix is a look-up table in which scores for each amino acid substitution have been calculated based on the frequency of that substitution in closely related protein's that have experienced a certain amount (x) of evolutionary divergence.

Phenotype: The observable characteristics of an organism, the expression of gene alleles (genotype) as an observable physical or biochemical trait. See Genotype.

Pheromone: A hormone-like substance that is secreted into the environment.

Phosphatase: An enzyme that hydrolyzes esters of phosphoric acid, removing a phosphate group.

Phosphodiester bond: A bond in which a phosphate group joins adjacent carbons through ester linkages. A condensation reaction between adjacent nucleotides results in a phosphodiester bond between 3' and 5' carbons in DNA and RNA.

Phospholipid: A class of lipid molecules in which a phosphate group is linked to glycerol and two fatty acyl groups. A chief component of biological membranes. (See Inositol phospholipid.)

Phosphorylation: The addition of a phosphate group to a compound.

Physical map: A map showing physical locations on a DNA molecule, such as restriction sites, and sequence-tagged sites. (See Mapping.)

Piezoelectric material: A ferroelectric material in which an electrical potential difference is created due to mechanical deformation, or conversely, in which the application of a voltage causes dimensional changes in the material.

Piezoelectric: A piezoelectric material expands or contracts when a voltage is applied across it. Piezoelectric tubes are often used to move the tip in an AFM or STM. If a piezoelectric crystal is compressed, a voltage appears across it. This effect is sometimes used to measure displacements. A commonly used piezoelectric is PZT (Pb,Zr)TiO3.

Plaque: A clear spot on a lawn of bacteria or cultured cells where cells have been Iysed by viral infection.

Polylinker: A short DNA sequence containing several restriction enzyme recognition sites that is contained in cloning vectors.

Polymorphisms: Variant forms of a particular gene that occur simultaneously in a population.

Polynucleotide: A DNA polymer composed of multiple nucleotides. (See Nucleotide.)

Polypeptide (protein): A polymer composed of multiple amino acid units linked by peptide bonds.

Primary cell: A cell or cell line taken directly from a living organism, which is not immortalized.

Primer: A short DNA or RNA fragment annealed to single-stranded DNA, from which DNA polymerase extends a new DNA strand to produce a duplex molecule.

Probe: A single-stranded DNA that has been radioactively labeled and is used to identify complementary sequences in genes or DNA fragments of interest. (See Multilocus probe.) A fragment of DNA, usually labeled with radioactive 32P, that detects homologous sequences on southern blots.

Prokaryote: A bacterial cell lacking a true nucleus; its DNA is usually in one long strand. See Eukaryote.

Promoter: A region of DNA extending 150-300 bp upstream from the transcription start site that contains binding sites for RNA polymerase and a number of proteins that regulate the rate of transcription of the adjacent gene. (See Constitutive promoter.)

Restriction enzyme: A class of enzymes that cut DNA at specific sequences called restriction sites. Restriction enzymes were key to making genetic engineering possible. See RFLP.

Restriction-fragment-length polymorphism (RFLP): Differences in nucleotide sequence between alleles at a chromosomal locus result in restriction fragments of varying lengths detected by Southern analysis.

Retrovirus: A member of a class of RNA viruses that utilizes the enzyme reverse transcriptase to reverse copy its genome into a DNA intermediate, which integrates into the hostcell chromosome. Many naturally occurring cancers of vertebrate animals are caused by retroviruses.

Reverse genetics: Using linkage analysis and polymorphic markers to isolate a disease gene in the absence of a known metabolic defect, then using the DNA sequence of the cloned gene to predict the amino acid sequence of its encoded protein.

Reverse transcriptase (RNA-dependent DNA polymerase): An enzyme isolated from retrovirus-infected cells that synthesizes a complementary (c)DNA strand from an RNA template.

RFLP: Restriction fragment length polymorphism. RFLPs are generated by digesting DNA with restriction enzymes. The DNA is separated by size by gel electrophoresis. Slight differences in homologous fragments may exist between individuals. These length polymorphisms are RFLPs. See blot.

Ribosome: Cellular organelle that is the site of protein synthesis during translation. See Organelle, Translation.

Ribosome-binding site: The region of an mRNA molecule that binds the ribosome to initiate translation.

Risk assessment: Defined as a formalized basis for the objective evaluation of risk in a manner in which assumptions and uncertainties are clearly considered and presented.

RNA (ribonucleic acid): An organic acid composed of repeating nucleotide units of adenine, guanine, cytosine, and uracil, whose ribose components are linked by phosphodiester bonds. (See Antisense RNA, Heterogeneous nuclear RNA, Messenger RNA,

Ribosomal RNA, RNA polymerase, Small nuclear RNA, Transfer RNA.)

RNA polymerase: Transcribes RNA from a DNA template. (See Polymerase, RNA.)

RNA: Ribonucleic acid. The three basic types are messenger RNA (mRNA), ribosomal RNA (rRNA), and transfer RNA (tRNA).

Roundup Ready: A trademark for plants that are genetically engineered to be resistant to the herbicide Roundup (technical name: glyphosate).

Salmonella: A genus of rod-shaped, gram-negative bacteria that are a common cause of food poisoning.

Satellite RNA (viroids): A small, self-splicing RNA molecule that accompanies several plant viruses, including tobacco ringspot virus.

Scanning Near Field Optical Microscopy: A method for observing local optical properties of a surface that can be smaller than the wavelength of the light used. A type of scanning probe microscopy that maps the near-field optical properties of an interface.

Scanning Probe Microscopy (SPM): A method for imaging nanoscale features of surfaces by scanning a sensor (probe) over a surface. Near-field effects such as tunneling, van der Waals forces, local fields and more are serially detected at localized points on the surface and used to create an SPM image.

Scanning Thermal Microscopy (SThM): A type of scanning probe microscopy that maps the local temperature and thermal conductivity of an interface. A method for observing local temperatures and temperature gradients on a surface. [NTN]

Scanning Tunneling Microscope (STM): An instrument able to image conducting surfaces to atomic accuracy; has been used to pin: molecules to a surface. A device in which a sharp conductive tip is moved across a conductive surface close enough to permit a substantial tunneling current (typically a nanometer or less). In a common mode of operation, the voltage is kept constant and the current is monitored and kept constant by controlling the height of the tip above the surface; the result, under favorable conditions, is an atomic-resolution map of the surface reflecting a combination of topography and electronic properties. The STM has been used to manipulate atoms and molecules on surfaces. A type of scanning probe microscopy that maps the local electrical properties of an interface.

Schwann cell: Glial cell responsible for forming myelin sheaths in the peripheral nervous system.

Science: The process of developing a systematized knowledge of the world through the variation and testing of hypotheses. The pursuit of knowledge and understanding, from the Latin term *scientia*, which means 'knowledge'.

Selectable marker genes: Genetic engineers must be able to select cells that have received the transgene from those that have not. Selecting out transgenic cells is done by co-transforming the cells with the transgene plus an additional gene called a selectable marker gene. Selectable marker genes are genes that encode easily detectable traits making transgenic cells discernible from non-transgenic cells. The two most commonly used selectable marker genes encode the traits of herbicide and antibiotic resistance.

Self-pollination: Pollen of one plant is transferred to the female part of the same plant or another plant with the same genetic makeup.

Semiconservative replication: During DNA duplication, each strand of a parent DNA molecule is a template for the synthesis of its new complementary strand. Thus, one half of a preexisting DNA molecule is conserved during each round of replication.

Sequence hypothesis: Francis Crick's seminal concept that genetic information exists as a linear DNA code; DNA and protein sequence are colinear.

Sequence: Noun: the particular order of nucleotides in a DNA or RNA fragment. Verb: to determine the particular order of nucleotides in a strand of DNA.

Sequence-tagged site (STS): A unique (single-copy) DNA sequence used as a mapping landmark on a chromosome.

Sexual reproduction: The process where two cells (gametes) fuse to form one hybrid, fertilized cell. See Asexual reproduction, Gamete, Hybrid.

Signal transduction: The biochemical events that conduct the signal of a hormone or growth factor from the cell exterior, through the cell membrane, and into the cytoplasm. This involves a number of molecules, including receptors, pro- teins, and messengers.

Site-directed mutagenesis: The process of introducing spe- cific base-pair mutations into a gene.

Small nuclear RNA (snRNA): Short RNA transcripts of 100-300 bp that associate with proteins to form small nuclear ribonucleoprotein particles (snRNPs), which participate in RNA processing.

Somatic cell gene therapy: The repair or replacement of a defective gene within somatic tissue.

Somatic cell: Any nongerm cell that composes the body of an organism and which possesses a set of multiploid chromosomes (diploid in most organisms). (See Gamete, Somatic cell gene therapy.)

Southern hybridization (Southern blotting): A procedure in which DNA restriction fragments are transferred from an agarose gel to a nitrocellulose filter, where the denatured DNA is then hybridized to a radioactive probe (blotting). (See Hybridization.)

Species: A classification of related organisms that can freely interbreed.

Spore: A form taken by certain microbes that enables them to exist in a dormant stage. It is an asexual reproductive cell. See Asexual reproduction, Dormant.

Stationary phase: The plateau of the growth curve after log growth, during which cell number remains constant. New cells are produced at the same rate as older cells die. (See Growth phase.)

Sticky end: A protruding, single-stranded nucleotide sequence produced when a restriction endonuclease cleaves off center in its recognition sequence. After cutting with restriction enzymes, the ends of DNA fragments are "sticky". They can easily be joined with other fragments that were cut with the same enzyme and so have complementary sticky ends.

Stringency: Reaction conditions—notably temperature, salt, and pH—that dictate the annealing of single-stranded DNA/DNA, DNA/RNA, and RNA/RNA hybrids. At high stringency, duplexes form only between strands with perfect one-to-one complementarity; lower stringency allows annealing between strands with some degree of mismatch between bases.

Stringent plasmid: A plasmid that only replicates along

with the main bacterial chromosome and is present as a single copy, or at most several copies, per cell. (See plasmid.)

Structure-functionalism: The scientific tradition that stresses the relationship between a physical structure and its function, for example, the related disciplines of anatomy and physiology.

Transfection: The uptake and expression of a foreign DNA sequence by cultured eukaryotic cells.

Transformant: In prokaryotes, a cell that has been genetically altered through the uptake of foreign DNA. In higher eukaryotes, a cultured cell that has acquired a malignant phenotype. (See Transformation.)

Transformation: (1) Transformation is the step in the genetic engineering process where a new gene (transgene) is delivered into the nucleus of a plant cell and inserts into a chromosome where it is passed on to progeny. Transformation means to genetically change a living thing. (2) In prokaryotes, the natural or induced uptake and expression of a foreign DNA sequence—typically a recombinant plasmid in experimental systems. In higher eukaryotes, the conversion of cultured cells to a malignant phenotype—typically through infection by a tumor virus or transfection with an oncogene. (See Transformant, Transformation efficiency.)

Transformation efficiency: The number of bacterial cells that uptake and express plasmid DNA divided by the mass of plasmid used (in transformants/microgram). (See Transformation.)

Transforming oncogene: A gene that upon transfection converts a previously immortalized cell to the malignant phenotype. (See Oncogene.)

Transgene: A foreign gene incorporated by transformation.

Upload: (a) To transfer the consciousness and mental structure of a person from a biological matrix to an electronic or informational matrix (this assumes that the strong AI postulate holds). The term "Downloading" is also sometimes used, mainly to denote transferring the mind to a slower or less spacious matrix. (b) The resulting infomorph person. [The origin of the term is uncertain, but obviously based on the computer technology term 'uploading' (loading data into a mainframe computer).] [AS]

Upstream: The region extending in a 5' direction from a gene.

Vaccine: A preparation of dead or weakened pathogen, or of derived antigenic determinants, that is used to induce formation of antibodies or immunity against the pathogen. (See Polyvalent vaccine, Subunit vaccine.)

Vaccinia: The cowpox virus used to vaccinate against smallpox and, experimentally, as a carrier of genes for antigenic determinants cloned from other disease organisms.

Variable surface glycoprotein (VSG): One of a battery of antigenic determinants expressed by a microorganism to elude immune detection.

Variation: Differences in the frequency of genes and traits among individual organisms within a population.

Vasculocyte: In medical nanorobotics, a theorized (nanorobotic) device capable of performing repairs of an injured vascular luminal surface.

Vasculography: A physical description (and mapping) of the human vascular system.

Vector: In cell biology, an agent (virus or plasmid) used to transmit genetic material to a cell or organism.

Vegetal pole: The end at which most of the yolk is located in an animal egg. The end opposite the animal pole.

Viral oncogene: A viral gene that contributes to malignancies in vertebrate hosts.

Viroid: A plant pathogen that consists of a naked RNA molecule of approximately 250-350 nucleotides, whose extensive base pairing results in a nearly correct double helix. (See Satellite RNA.)

Virtual Nanomedicine: Using VR to perform surgery and other functions inside the body.

Virtual reality system: A combination of computer and interface devices (goggles, gloves, etc.) that presents a user with the illusion of being in a three dimensional world of computer-generated objects.

1

Biotech Engineering: An Introductory Overview

Bioengineering is the branch of engineering science in which biological science is used to study the relation between workers and their environments. It deals with the application of engineering principles to the fields of biology and medicine, as in the development of aids or replacements for defective or missing body organs and is also called biomedical engineering.

Protein engineering is the application of science, mathematics, and economics to the process of developing useful or valuable proteins. It is a young discipline, with much research currently taking place into the understanding of protein folding and protein recognition for protein design principles. There are two general strategies for protein engineering. The first is known as *rational design*, in which the scientist uses detailed knowledge of the structure and function of the protein to make desired changes. This has the advantage of being generally inexpensive and easy, since site-directed mutagenesis techniques are well-developed. However, there is a major drawback in that detailed structural knowledge of a protein is often unavailable, and even when it is available, it can be extremely difficult to predict the effects of various mutations. Computational protein design algorithms seek to identify amino acid sequences that have low energies for target structures. While the sequence-conformation space that needs to be searched is large, the most challenging requirement for computational protein design is a fast, yet accurate, energy function that can distinguish optimal sequences from similar suboptimal ones. Using computational methods, a protein with a novel fold has been designed, as well as sensors for un-natural molecules. The second strategy is known as directed evolution.

This is where random mutagenesis is applied to a protein, and a selection regime is used to pick out variants that have the desired qualities. Further rounds of mutation and selection are then applied. This method mimics natural evolution and generally produces superior results to rational design. An additional technique known as DNA shuffling mixes and matches pieces of successful variants in order to produce better results.

This process mimics recombination that occurs naturally during sexual reproduction. The great advantage of directed evolution techniques is that they require no prior structural knowledge of a protein, nor it is necessary to be able to predict what effect a given mutation will have. Indeed, the results of directed evolution experiments are often surprising in that desired changes are often caused by mutations that no one would have expected. The drawback is that they require high-throughput, which is not feasible for all proteins. Large amounts of recombinant DNA must be mutated and the products screened for desired qualities.

The sheer number of variants often requires expensive robotic equipment to automate the process. Furthermore, not all desired activities can be easily screened for. Rational design and directed evolution techniques are not mutally exclusive; good researchers will often apply both. In the future, more detailed knowledge of protein structure and function, as well as advancements in high-throughput technology, will greatly expand the capabilities of protein engineering.

Bioengineered Foods

Very basically, food-related biotechnology is the process by which a specific gene or group of genes with desirable traits are removed from the DNA of one plant or animal cell and spliced into that of another. Such beneficial genes might come from animals, (friendly) bacteria, fish, insects, plants and even humans.

In some instances, genes that create problems (such as the natural softening of a tomato) are simply removed and not replaced. Tomatoes, for example, are generally picked green and gas-ripened later because, during shipping, they would become soft, bruised and unmarketable. A bioengineered tomato, however, can be picked ripe and shipped without softening.

The objective of food biotechnology is to develop insect- and disease-resistant, shipping- and shelf-stable foods with improved appearance, texture and flavor. Additionally, biotechnology advocates say that the process will produce plants that are resistant to adverse weather conditions such as drought and frost, thereby increasing food production in previously prohibitive climate and soil conditions. They also envision increasing nutrient levels and decreasing pesticide usage through biotechnology.

On the other hand, critics argue that, because biotechnology is producing new foods not previously consumed by humans, the changes and potential risks relating to such things as toxins, allergens and reduced nutrients are unpredictable. They also worry that, because genetically altered foods are not required to be labeled, people with religious or lifestyle dietary restrictions might unintentionally consume prohibited foods. In answer to such concerns, the FDA has issued the following evaluation guidelines by which a bioengineered food will be judged for approval:

- Has the concentration of a plant's naturally occurring toxicant increased?
- Has an allergic element not commonly found in the plant been introduced?
- Have the levels of important nutrients changed?
- Have accepted, established scientific practices been followed?
- What are the effects on the environment?

Genetic engineering is the process of manually adding new DNA to an organism. The goal is to add one or more new traits that are not already found in that organism. Examples of genetically engineered (transgenic) organisms currently on the market include plants with resistance to some insects, plants that can tolerate herbicides, and crops with modified oil content. To understand how genetic engineering works, there are a few key biology concepts that must be understood DNA is the recipe for life. DNA is a molecule found in the nucleus of every cell and is made up of 4 subunits represented by the letters A, T, G, and C.

The order of these subunits in the DNA strand holds a code of information for the cell. Just like the English alphabet makes up words using 26 letters, the genetic language uses 4 letters to spell out the instructions for how to make the proteins an organism will need to grow and live. Small segments of DNA are called genes. Each gene holds the instructions for how to produce a single protein. This can be compared to a recipe for making a food dish. A recipe is a set of instructions for making a single dish. An organism may have thousands of genes. The set of all genes in an organism is called a genome. A genome can be compared to a cookbook of recipes that makes that organism what it is. Every cell of every living organism has a cookbook.

Proteins do the work in cells. They can be part of structures (such as cell walls, organelles, etc). They can regulate reactions that take place in the cell. Or they can serve as enzymes, which speed-up reactions. Everything you. see in an organism is either made of proteins or the result of a protein action Small segments of DNA are called genes. Each gene holds the instructions for how to produce a single protein. This can be compared to a recipe for making a food dish.

A recipe is a set of instructions for making a single dish. An organism may have thousands of genes. The set of all genes in an organism is called a genome. A genome can be compared to a cookbook of recipes that makes that organism what it is. Every cell of every living organism has a cookbook. DNA is a 'universal language', meaning the genetic code means the same thing in all organisms.

It would be like if all cookbooks around the world were written in a single language that everyone knew. This characteristic is critical to the success of genetic engineering. When a gene for a desirable trait is taken from one organism and inserted into another, it gives the 'recipient' organism the ability to express that same trait. Genetic engineering, also called transformation, works by physically removing a gene from one organism and inserting it into another, giving it the ability to express the trait encoded by that gene. It is like taking a single recipe out of a cookbook and placing it into another cookbook.

The Process: Once a Goal is in Mind...

1. First, find an organism that naturally contains the desired trait.
2. The DNA is extracted from that organism. This is like taking out the entire cookbook.
3. The one desired gene (recipe) must be located and copied from thousands of genes that were extracted. This is called gene cloning.
4. The gene may be modified slightly to work in a more desirable way once inside the recipient organism.
5. The new gene(s), called a transgene is delivered into cells of the recipient organism. This is called transformation. The most common transformation technique uses a bacteria that naturally genetically engineer plants with its own DNA. The transgene is inserted into the bacteria, which then delivers it into cells of the organism being engineered. Another technique, called the gene gun method, shoots microscopic gold particles coated with copies of the transgene into cells of the recipient organism. With either technique, genetic engineers have no control over where or if the transgene inserts into the genome. As a result, it takes hundreds of attempts to achieve just a few transgenic organisms.
6. Once a transgenic organism has been created, traditional breeding is used to improve the characteristics of the final product. So genetic engineering does not eliminate the need for traditional breeding. It is simply a way to add new traits to the pool.

Although the goal of both genetic engineering and traditional plant breeding is to improve an organism's traits, there are some key differences between them. While genetic engineering manually moves genes from one organism to another, traditional breeding moves genes through mating, or crossing, the organisms in hopes of obtaining offspring with the desired combination of traits. Using the recipe analogy, traditional breeding is like taking two cookbooks and combining every other recipe from each into one cookbook. The product is a new cookbook with half of the recipes from each original book. Therefore, half of the genes in the offspring of a cross come from each parent. Traditional breeding is effective in improving traits, however, when compared with genetic engineering, it does have disadvantages.

Since breeding relies on the ability to mate two organisms to move genes, trait improvement is basically limited to those traits that already exist within that species. Genetic engineering, on the other hand, physically removes the genes from one organism and places them into the other. This eliminates the need for mating and allows the movement of genes between organisms of any species. Therefore, the potential traits that can be used are virtually unlimited. Breeding is also less precise than genetic engineering. In breeding, half of the genes from each parent are passed on to the offspring. This may include many undesirable genes for traits that are not wanted in the new organism. Genetic engineering, however, allows for the movement of a single, or a few, genes.

During the past decade, biotechnology companies commercialized the first generation of genetically engineered crops—primarily corn, soybeans, and cotton altered to control insects and weeds. U.S. commodity crop producers responded by planting millions of acres of these engineered crops. Because corn and soy are widely used in food processing, small amounts of engineered ingredients show up in a majority of processed food products. But most foods—the vast majority of vegetables, grains, fruits, and nuts—remain unaltered. Of the eight other engineered food plants allowed in U.S. grocery stores, it appears that only engineered canola and papaya are currently available. Among food animals, only engineered fish are under active consideration by U.S. regulators. Other engineered plants, animals, and microbes are farther down the research pipeline but few are poised for introduction in the near future. Recently, a second wave of biotech products began emerging—crops, mainly corn, engineered to produce pharmaceuticals and industrial and research chemicals.

Several such products are already on the market and companies are seeking approval from the Food and Drug Administration of corn-based drugs and vaccines. Scientists are concerned that engineered organisms might harm people's health or the environment. For example, engineered crops might contaminate the food supply with drugs, kill beneficial insects, or jeopardize valuable natural resources like Bt toxins. Engineered fish may substantially alter native ecosystems, perhaps even driving wild populations to extinction. To protect human health and the environment from engineered products, we need strong federal oversight and active citizen participation. We urge you to join in our efforts to strengthen U.S. regulation of agricultural biotechnology products. Our current priorities are to:

- Convince the federal government to establish regulations to protect the food supply and environment from contamination by engineered pharma and industrial crops
- Persuade the Environmental Protection Agency to conduct rigorous reviews of ecological risks and require strong resistance-management plans before approving crops producing Bt toxins
- Press the federal government to strengthen its

oversight of the environmental risks of engineered fish

- Urge the Food and Drug Administration to require safety testing and labeling before biotech foods are allowed on the market.

Developments Genetic Engineering in USA and UK

The Biotechnology Information Center (BIC) is one of ten information centres at the National Agricultural Library located in Beltsville, Maryland. The Biotechnology Information Center provides access to a variety of information services and publications covering many aspects of agricultural and environmental biotechnology, including transgenic animals, current gene mapping projects, legislation and regulation, patenting issues and biotechnology newsletters. There are also some collections of resources on more specific topics, such as bovine somatotropin (bST/BGH), bioremediation and Bacillus thuringiensis.

It should be noted that the Biotechnology Resource Web site is no longer updated and is useful for general purposes only. Humanisation (also called CDR-grafting or Reshaping) is a technique to reduce the immunogenicity of monoclonal antibodies (mAbs) from xenogenic sources, improving their activation of the human immune system. This site is about the design of the engineered mAb, supplying data and raising design issues to help the prospective antibody designer. Selected humanised antibodies from the literature can be searched by text or sequence. Historical details are also available, an introduction to antibody structure, and a list of useful links. It is produced by Jose Saldanha of the UK National Institute for Medical Research, and made available on the Web by the Department of Crystallography Birkbeck College, London. The National Centre for Biotechnology Education (NCBE) was founded in 1985, as part of the Department of Microbiology at the University of Reading. It aims to support the teaching of biotechnology in schools, providing training, resources and information for schools, industry, the public, and professional organisations. Information is provided here about events, materials, reviews of resources, protocols, safety information, technical advice, and links to other relevant resources.

Scientific Advisory Committee on Genetic Modification

Scientific Advisory Committee on Genetic Modification, an advisory committee of the Health & Safety Commission. The role of the SACGM is to "advise the Health and Safety Commission and Executive and the Secretary of State for the Environment, and other Ministers and bodies, as appropriate, on all aspects of the human and environmental safety of the contained use of genetically modified organisms." Full-text of agendas and newsletters are provided here, as well as information on the contained use of genetically modified organisms. Made available on the Web by the UK government's Health and Safety Commission. A report published by the Global Change Development Programme. This report addresses the issue of genetically modified food, examining the politics, ethics and scientific thinking surrounding its implementation in the UK. The emphasis of the report is focused on the UK government's handling of the GM food debate. References are provided and many are available via the Web. The report can be downloaded free in either PDF (requiring Adobe Acrobat Reader software) or HTML. The report is aimed at anyone with an interest in the UK GM food debate. This was a ten year research programme established in 1991, and was funded by the Economic and Social Research Council. A report from a Working Party at the Nuffield Council on Bioethics, on the ethical and social issues surrounding GM technology. Chapters are included on the scientific and ethical principles of genetic technologies, the scientific basis of genetic modification, issues related to commercial implementation, the impact on developing countries, consumer choice and food quality, the environmental impact, policy, and conclusions and recommendations. Published in May 1999, the report is provided in HTML format and in PDF (requiring Adobe Acrobat Reader).

Biosafety Information Network and Advisory Service (BINAS): It is a service of the United Nations Industrial Development Organization (UNIDO). BINAS is responsible for monitoring global developments in regulatory issues in biotechnology. This site provides the BINASNews newsletter in full-text, which covers articles on topics such as genetically modified food and crops, and other socioeconomic issues within biotechnology.

The UK Advisory Committee on Releases to the Environment (ACRE): It is part of the governments Department of the Environment, Transport and the Regions (DETR). ACRE gives advice to the government on issues concerning human and environmental safety concerning the release of genetically modified organisms (GMOs) and non-native organisms into the environment. It is provided by the Biotechnology Unit of the DETR, which is responsible for the operation and policy of the legislation controlling GMO release. This site provides the Committees agendas and reports, full-text newsletters, summary records for experimental releases and for marketing GMOs in Europe, information on international agreements, press releases and advice to ministers. Details of the Advisory Committee on Genetic Modification can also be found here. Published on the Web by the DETR. GeneWatch UK is a non-profit policy research and public interest group. The areas of interest are divided into the following sections: genetically modified (GM) crops and food, human genetics, GM animals, laboratory use of GMOs, patenting, and bio-weapons. A comprehensive archive of publications relating to these topics is provided. A database of information on GM crops and food is also available, which was developed in assistance with a grant from the Greenpeace Environmental Trust. GeneWatch is aimed at anyone who has an interest in the ethical issues raised by genetic engineering. This Web site has been included in the Biotechnology section of the Nature Net Guide as an example of "what impressive things can be done in communicating and facilitating research". The Council for Responsible Genetics (CRG), established in 1983, is a US non-profit organisation of scientists, public health advocates, and others. Their main aim is "to promote a comprehensive public interest agenda for biotechnology." Online information available on this site includes genetic discrimination, cloning, patenting of life forms, and genetically engineered food. GeneWatch is a bi-monthly bulletin dedicated to the social and environmental implications of biotechnology, and selected articles are available online from January 1999. Other resources which are found on this site are CRG alerts on biotechnology development and an online petition. This site is aimed at anyone who is interested in the impact of new biotechnologies and their sociological and environmental impact. The International Centre for Genetic Engineering and Biotechnology (ICGEB) is a statutory body founded in 1983 through an initiative led by the United Nations Industrial Development Organization (UNIDO). In the beginning ICGEB operated as a special programme of UNIDO, gaining status in 1995 as an independent, fully autonomous international, intergovernmental organisation. The major goal of the ICGEB is to advance research and training in molecular biology and biotechnology, particularly in the developing world. ICGEB operates from two main centres, Trieste, Italy and New Delhi, India. This Web site provides details on affiliated centres (46 countries are currently full members), continuing research and ongoing activities, fellowships, grants and sponsored training courses, and web resources for biosafety, biotechnology transfer, patents and a computer resource for molecular biology called ICGEBnet. This leaflet summarises the controls on the contained use of genetically modified organisms, set out in the Genetically Modified Organisms (Contained Use) Regulations 2000. These Regulations are supplemented by the sections of the Environmental Protection Act 1990 and the Genetically Modified Organisms (Risk Assessment) (Records and Exemptions) Regulations (as amended in 1997) which specifically cover the control of risks to the environment from genetically modified animals and plants. The leaflet lists what the regulations cover, what is meant by contained use, what the main duties are under the regulations, who enforces the regulations, how the work is classified, notifications and consents, how notification should be made, and fees. It also outlines in brief the responsibilities of other Government organisations dealing with biotechnology, giving their addresses and telephone numbers for further information. This leaflet is published by the Health and Safety Executive and was added to the Web in November 2000. Published on the Web by the Biotechnology and Biological Sciences Research Council (BBSRC), this paper was authored by Dr. Robert Straughan in September 2000. This document "examines the ethical, moral and social issues surrounding relatively recent developments that involve the genetic modification of animals, and cloning by nuclear transfer." Sections include: Current developments in animal biotechnology; Moral and ethical concerns; Animal ethics; Intrinsic concerns aabout animal technology; Entrinsic concerns aabout animal technology; and Conclusions. A technical annex is provided as are bibliographic references.

MICER (Mutagenic Insertion and Chromosome Engineering Resource) is a tool for the targeted modification of the mouse genome. It is based on the mouse genome sequence: randomly defined sections of the mouse genome have been joined to a DNA molecule that can disrupt normal gene activity. This resource provides vector maps and sequences and embryonic stem cell protocols for using MICER vectors for generating knockout mice, and for chromosome engineering. A Sanger Clone Request form is available for researchers to obtain MICER vectors. Made available on the Web by the Wellcome Trust Sanger Institute. The Aptamer Database is designed to contain comprehensive sequence information on aptamers and unnatural ribozymes that have been generated by in vitro selection methods. Such data are not normally collected in sequence databases such as GenBank. Besides serving as a repository of sequences that may have diagnostic or therapeutic utility, the database serves as a valuable resource for theoretical biologists who describe and explore fitness landscapes. The database is updated monthly. Made available on the Web by the Institute for Cellular and Molecular Biology, University of Texas at Austin.

Economic Impact of Genetic Engineering on Developing Countries

Genetic engineering (GE) is a recently developed technology that allows the alteration of the genetic make up of living organisms. This technology allows scientists not only to exchange genes from members of the same species, which is what farmers and nature has been doing through out history, but also the exchange of genes between completely separate species. For example genetic engendering allows scientists to insert the genes from a fish into a tomato, something that can never happen in nature. Many in the biotechnology industry claim that the recent advancements in this field of research will help solve many of the problems developing countries face. These claims have led to a massive increase in commercial growing of genetically modified organisms, (GMOs) from 1.7 million hectares in 1996 to 58.1 million hectares in 2002 worldwide. 99% of this growth has only been in four countries; Argentina, Canada, China and the USA, the USA being the largest, producing 68%. Amidst this enthusiasm there has been growing concern that there exist many hidden costs involved with GE technology. Critics of GE technology claim there are known and unknown health, social, ethical and environmental risks involved with the introduction of this technology. There are also claims that this technology has little if any benefits to the developing world and will in fact only benefit the developed world, and multinational biotech corporations such as Monsanto. Critics argue that whatever benefits developing nations gain from GE technology, these are far outweighed by the risks involved. Proponents' of genetically modified foods strongest claim is that this technology will benefit farmers in developing countries by allowing those farmers to be more productive and grow a higher yield of crops. Many studies do agree that GE technologies do give farmers a higher yield, but even this is a contested issue. Because many of these studies are sponsored and linked to the very corporations that are developing these technologies, validity of the results are questionable. In fact, there have been certain independent studies that have concluded that genetically modified seeds in fact produce a lower yield than do conventional seeds. For example these is a two-year study by the University of Nebraska, in which Monsanto's Roundup Ready soybeans yield 6 percent less than their closest relatives and 11 percent less than high-yielding soybean varieties. This study shows that with all things being equal, genetically modified crops may in fact produce a lower yield. Because GE soy is commercially grown more than any other GE crop it will be good to take a look at why. With Roundup Ready soybeans growing to 81% of the totally soy grown within the United States in 2003 , one must ask why farmers are switching to GE soy. The reason for this is the lower costs of pesticides and herbicides. The advantage of Roundup Ready soybeans is not that they are more productive or that the seeds cost less. In fact the seeds cost about six dollars more per acre than traditional soy seeds. The advantage comes in the form of a more resistant plant. Because Roundup Ready soy has been genetically altered to be more resistant to herbicides and pesticides, specifically a Monsanto produced brand called Roundup. Because these new GE soybeans are able to withstand stronger herbicide and pesticide use at a younger age, without any damage to the crop, farmers are able to kill off weeds and pests earlier and easier. This use of stronger herbicides and pesticides in farming leads to a cleaner crop, with a lower cost of labor. A 1998 study by Mike Duffy, an economist from Iowa State University shows that all in all the costs and benefits of using GE technologies in soy and corn production do not lead to any sort of significant difference in returns compared to traditional none GE seeds. The main reasons for farmers switching to genetically altered technologies is the lower amount of labor needed for production and the cleaner crop produced with an increase of herbicides. Clearly the fact that Roundup Ready soy, the most highly used commercial GMO product, has lower yields than traditional soy products we can dismiss the main argument given by the biotech industry that they are trying to help end world hunger. According to the United Nations and many none governmental organizations around the world the problem of world hunger is not one of a scarcity of food; it is in fact a problem of distribution of food. Today the world produces 17 percent more calories per person today than it did 30 years ago . The world does not need larger supplies of food; the world needs to address the way in which food is distributed, something that GMO technology has no effects on. The only people that the biotech industry is trying to help are themselves. This is the reason that currently the main commercial implementation of genetically modified technology has been to increase crop resistance to pesticide use. The main producers of GE technology, firms such as Monsanto, are also the world's main producers of pesticides and herbicides. The example of soy shows that the only real impact of genetic engineering has been to transfer some of the cost of labor into capital costs. This may be a good thing for the developing world where labor is more scarce, and capital intensive production should be sought. The developing world, on the other hand, with its high population and high unemployment would be better off using more labor intensive means of production. If GE soy technologies are applied in less developed nations, where labor is cheaper than in the developed countries, the cost of capital will not be offset as it has been in the United States by lowering labor costs. The costs of GE seeds are too high for farmers from developing countries to implement their usage. The costs of implementing this technology will not be offset by the benefits that farmers in the US have received. Not only will farmers lose money, but unemployment, which is already a problem in many developing nations, will increase, as farmers need to hire less labor to produce the same amount of crops. The high initial cost of purchasing GE seeds makes the technology more difficult to implement by poor farmers in less developed countries. As happened during the "green revolution" these small farmers would be forced to take out loans to buy this new technology, and if the technology does not deliver higher profits these farmers may lose their land. Even if a technology was developed that would bring the cost of production down to or below that of existing production cost, there is the issue of consumer preference. The increasing debate over genetic engineering has created a substantial difference in perception of the risks and benefits involved with GE food. Governments around the world, most notably members of the European Union have places restrictions and bans on many GMO products. Many of these countries are considering putting labels on products with GMO ingredients. Considering that many developing countries' economies are highly dependant on agricultural export, these countries can not risk the loss of major export markets. With pressure from the biotech industry, the United States, and the World Trade Organization some countries are become more lax on their restrictions on GE imports and regulations. The most recent example is the lifting of a moratorium on commercial GMOs in New Zealand on October 29th, 2003. While the governments of many countries are slowly lowering their restrictions on GE technology, many of the consumers in these countries are still highly opposed to the consumption of GMOs. The highest consumer opposition to this technology comes from Western Europe and Japan. With many consumers in the developed world being highly opposed to GMOs and willing to pay a different, higher price for GMO free food, GE crops stop being a substitute for traditional crops, and two separate markets are formed, with different demand curves. Even if a GE technology is developed that allows higher productivity at equal or slightly lower costs, this may still not be a profitable investment for farmers that export to locations such as Western Europe where there is high consumer resistance to GMOs. In a market where GE foods are not substitutes for traditional foods, many consumers will be unwilling to purchase GE products. Also, if GE technology does become more productive than traditional food, those who are willing to substitute GE food for traditional food will have a larger supply of both genetically modified and traditional food to chose from.

With the lower demand, and higher supply, the price of genetically modified foods will fall. What benefits the farmers of GMOs received in switching to this technology will be accompanied by a loss of price of their goods. The option of growing a limited amount of genetically modified products, and keeping the rest of the market GMO free, is a very costly, if not impossible task. Fear of GMOs spreading to the wild and having negative effects on the local ecosystems is one of the main environmental arguments against GE technology. Some countries have attempted to stop this by creating "buffer zones" between genetically altered plants and traditional plants. The problem with this is most of these buffer zones are too small in size. Genetically altered plants have been found to spread their pollen over 800 meters , it is highly unlikely and would be extremely costly for farmers to have to implement buffer zones this large around their farms to make sure their crop does not contaminate any outside crops. While this is currently an argument environmentalists are using to show that GMOs are a threat to the ecosystem, this is not only a threat to the environment, but also to other farmers who wish to farm using traditional seeds. Farmers who try to sell their products to the none GMO market and have their fields contaminated by pollen from their neighbor's GE crop will not be able to sell their products as GMO free. If the market for GMO product goes bad it will be close to impossible for a country to stop production of GMOs once they get out of their contained areas.

One technological innovation that the biotech industry has come up with that does not have this problem is the terminator technology. Industry wide research has been devoted to expanding this technology. Terminator seeds are seeds which create plants that are sterile and cannot produce new seeds. With this technology farmers are forced to purchase new seeds every year from biotech corporations, unlike traditionally where farmers were able to use seeds from the previous harvest to plant the next year's crop. The reason for developing this technology is simply to protect the patent on the corporations genetically altered product. Forcing the purchase of new seeds every season would hurt farmers in poorer developing nations most. For generations farmers around the world have been harvesting seeds from their best plants every year, naturally evolving their crops to be better. The ability to breed the best plants with one another is taken away when farers plant sterile seeds. This technology will force farmers from developing nations to be more dependant on multinational biotech corporations. While terminator technology has been met with large criticisms and biotech firms have been slow to implement its uses in the United States, because of fears of their patents not being enforced in developing countries, to ensure that their technology is not used by anyone without their permission, biotech companies are more likely to use terminator technologies in developing countries. Critics of GMOs claim that the long term health effects of these technologies are not known yet. The economic risks involved are also not all known. "Super weeds" that are not stopped by herbicides can cause a lower crop yield. Many scientists argue that usage of higher amounts of herbicides will create resistance in certain types of weeds, and these "super weeds" will not be stopped by herbicide use. Another, more alarming trend is that certain traits of genetically altered crops have crossed into weeds and those weeds now posses the same resistance to herbicides that the genetic alterations gave the crop. Under monoculture a farmer's crops are at higher risk of being destroyed by a disease. Because a farmer grows all the same species, if a disease breaks out, that farmers whole crop can be ruined. This is the same danger with GMOs only to a higher extent. Because genetically modified crops are exactly the same genetically, the risk from diseases is much higher. Unlike traditional crops which may be divers enough to avoid a totally disaster, GE crops have no diversity. Not only is GE implementation a bad idea for farmers from developing countries, it is already threat to many farmers who produce things such as sugar. Biotech firms are altering the genetic makeup of certain crops grown in northern climates to produce sugar and other derivatives, traditionally only obtained from crops grown in developing countries. With the development of this technology developed countries in the north will be able to grow products that they otherwise would have had to trade for. For example, coconut oil is naturally high in lauric acid. However it has become threatened by genetically engineered high-lauric acid canola oil grown in developed countries. While genetic engineering has been attacked on environmental, health and ethical grounds, it is clear that economically this technology will harm the developing world more than it can ever help. The technology is not being used to feed the world's hungry, nor does it need to be, distribution is the problem. Through patens and higher use of their own pesticides, GE technology has so far only been used to make biotech companies more money. The costs of implementing this technology are too high. With a higher investment in capital, and unknown returns in not something a low income farmer can afford. This would be especially costly if terminator seeds are implemented in developing countries, due to fear of patents not being honored in those countries. Developing countries should focus on more labor intensive production that does not leave them dependant on multinational corporations and the developed world.

Applications of Nanotech

Laboratories throughout the world are rapidly gaining atomically precise control over matter. As this control extends to an ever wider variety of materials, processes and devices, opportunities for applications relevant to NASA's missions will be created. This document surveys a number of future molecular nanotechnology capabilities of aerospace interest. Computer applications, launch vehicle improvements, and active materials appear to be of particular interest. We also list a number of applications for each of NASA's enterprises. If advanced molecular nanotechnology can be developed, almost all of NASA's endeavors will be radically improved. In particular, a sufficiently advanced molecular nanotechnology can arguably bring large scale space colonization within our grasp. This study describes potential aerospace applications of molecular nanotechnology, defined as the thorough three-dimensional structural control of materials, processes and devices at the atomic scale. The inspiration for molecular nanotechnology comes from Richard P. Feynman's 1959 visionary talk at Caltech in which he said, "The problems of chemistry and biology can be greatly helped if our ability to see what we are doing, and to do things on an atomic level, is ultimately developed—a development which I think cannot be avoided." Indeed, scanning probe microscopes (SPMs) have already given us this ability in limited domains. Synthetic chemistry, biotechnology, "laser tweezers" and other developments are also bringing atomic precision to our endeavors.

An expanded version of Drexler's MIT PhD thesis, examines one vision of molecular nanotechnology in considerable technical detail. It proposes the development of programmable molecular assembler/replicators. These are atomically precise machines that can make and break chemical bonds using mechanosynthesis to produce a wide variety of products under software control, including copies of themselves. Interestingly, living cells exhibit many properties of assembler/replicators. Cells make a wide variety of products, including copies of themselves, and can be programmed with DNA. Replication is one approach to building large systems, such as human rated launch vehicles, from molecular machines manipulating matter one or a few atoms at a time. Note that biological replication is responsible for systems as large as redwood trees and whales. Another approach to nanotechnology is supramolecular self-assembly, where molecular systems are designed to attract each other in a particular orientation to form larger systems. Hollow spheres large enough to be visible in a standard light microscope have been created this way using self-assembling lipids. There are many other examples and this field is rapidly advancing. Biological systems can do most of what molecular nanotechnology strives to accomplish — atomically precise products, active materials, reproduction, etc. However, biological systems are extremely complex and molecular nanotechnology seeks simpler systems to understand, control and manufacture. Also, biological systems usually work at fairly mild temperature and pressure conditions in solution — conditions that are not found in most aerospace environments.

Today, extremely precise atomic and molecular manipulation is common in many laboratories around the world and our abilities are rapidly approaching Feynman's dream. The implications for aerospace development are profound and ubiquitous. A number of applications are mentioned here and a few are described in some detail with references. From this sample of applications it should be clear that although molecular nanotechnology is a long term, high risk project, the payoff is potentially enormous — vastly superior computers, aerospace transportation, sensors and other technologies; technologies that may enable large scale space exploration and colonization. This chapter is organized into two sections. In the first, we examine three technologies—computers, aerospace transportation, and active materials—useful to nearly all NASA missions. In the second, we investigate some potential molecular nanotechnology payoffs for each area identified in NASA's strategic plan. Some of these applications are under investigation by nanotechnology researchers at NASA Ames. Some of the applications described below have relatively near-term potential and working prototypes may be realized within three to five years. This is certainly not true in other cases. Indeed, many of the possible applications of nanotechnology that we describe here are, at the present time, rather speculative and futuristic.

However, each of these ideas have been examined at least cursorily by competent scientists, and as far as we know all of them are within the bounds of known physical laws. We are not suggesting that their achievement will be easy, cheap or near-term. Some may take decades to realize; some other ideas may be scrapped in the coming years as insuperable barriers are identified. But we feel that they are worth mentioning here as illustrations of the potential future impact of nanotechnology. The applicability of manufacturing at an ever smaller scale is nowhere more self-evident than in computer technology. Indeed, Moore's law (an observation not a physical law) says that computer chip feature size decreases exponentially with time, a trend that predicts atomically precise computers by about 2010-2015. This capability is being approached from many directions. Here we will concentrate on those under development by NASA Ames and her partners. Carbon nanotubes can be viewed as rolled up sheets of graphite from 0.7 to many nanometers in diameter. The smaller tubes are single molecules. Dai placed carbon nanotubes on an SPM tip thus extending our ability to manipulate a single molecule with sub-angstrom accuracy. Not only are the tips atomically precise, but they should have approximately the same chemistry as C_{60}, and thus be functionalizable with a wide variety of molecular fragments. Functionalizing carbon nanotube tips will allow mechanical manipulation of many molecular systems on various surfaces with sub-angstrom accuracy. One particularly intriguing possibility along this line is to utilize a carbon nanotube SPM tip to engrave patterns on a silicon surface. It should be possible to create features a few nanometers across. These would be perhaps 100 times finer than the current state of the art in commercial semiconductor photolithography.

Further, in contrast to approaches such as electron microscope lithography for which the speed of operation now appears to be an insuperable obstacle for industrial production, nanotube SPM-based lithography can be accelerated by utilizing an array with thousands of SPM tips simultaneously engraving different parts of a silicon surface. Also, nanotube SPM lithography could provide a practical means to explore various futuristic electronic device technology ideas, such as quantum cellular automata, which require exceedingly small feature sizes. Needless to say, if these ideas pan out, they could literally revolutionize computer device technology, paving the way for systems that are many times more powerful and more compact than any available today. For the near term, it should be noted that the semiconductor industry is a major market for SPM products. These are used to examine production equipment. High performance carbon nanotube tips should be of substantial value. NASA Ames is collaborating with Dr. Dai, now at Stanford, to develop these tips. It is possible to store data on long chain molecules (for example, DNA) and it may be possible to read these data with carbon nanotube tipped SPMs. Existing DNA synthesis techniques can be used to write data. If the different DNA base pairs can be distinguished with a carbon nanotube tipped SPM, then the data can be read non-destructively. However, the difference between base pairs is not great. If the base pairs cannot be distinguished, techniques for attaching modified enzymes to specific base pair sequences could be used. Certain enzymes (DNA (cytosine-5) methyltransferases) attach themselves onto a specific sequence of base pairs with a covalent bond. The enzyme then performs its operation and breaks the bond. Smith modified the enzyme such that the initial covalent bond was formed but the subsequent operation was disrupted. The result is that DNA synthesized with the target base pair sequences at the desired location can force precise placement of the enzymes. The presence of an enzyme could represent 1 and its absence 0. Enzymes are sufficiently large that distinguishing their presence should be straightforward. If the DNA/enzyme approach proves impossible, a wide variety of other polymer systems could be examined. Bauschlicher computationally studied storing data in a pattern of fluorine and hydrogen atoms on the diamond surface. If write-once data could be stored this way, 10^{15} bytes/cm^2 is theoretically possible. By comparison, the new DVD write-once disks now coming on the market hold about 10^8 bytes/cm^2. Bauschlicher compared the interaction of different probe molecules with a one dimensional model of the diamond surface. This study found some molecules whose interaction energies with H and F are sufficiently different that the force differential should be detectable by an SPM. These studies were extended to include a two dimensional model of the diamond surface and two other systems besides F/H. Other surfaces, such as Si, and other probes, such as those including transition metal atoms, have also been investigated. Among the better probes was C_5H_5N (pyridine).

Quantum calculations suggest that pyridine is stable when attached to C_{60} in the orientation necessary for sensing the difference between hydrogen and fluorine. Half of C_{60} can form the end cap of a (9,0) or (5,5) carbon nanotube, and carbon nanotubes have been attached to an SPM tip. Thus, it might be possible using today's technology to build a system to read the diamond memory surface. Avouris has shown that individual hydrogen atoms can be removed from a silicon surface. If this could be accomplished in a gas that donates fluorine to vacancies on a diamond surface, the data storage system could be built. Thummel computationally investigated methods for adding a fluorine at the radical sites where a hydrogen atom had been removed from a diamond surface. As mentioned before, carbon nanotubes can be described as rolled up sheets of graphite. Different tubes can have different helical windings depending on how the graphite sheet is connected to itself.

Theory suggests that single-walled carbon nanotubes can have metallic or semiconductor properties depending on the helical winding of the tube. Chico, Han, Menon, and others have computationally examined the properties of some of hypothetical devices that might be made by connecting tubes with different electrical properties. Such devices are only few nanometers across — 100 times smaller than current computer chip features. Several authors, including Tour have described methods to produce conjugated macromolecules of precise length and composition. This technique was used to produce molecular electronic devices in mole quantities. The resultant single molecular wires were tested experimentally and found to be conducting. The three and four terminal devices have been examined computationally and look promising. The features of these components are approximately 3 angstroms wide, about 750 times smaller than current silicon technology can produce.

Helical logic is a theoretical proposal for a future computing technology using the presence or absence of individual electrons (or holes) to encode 1s and 0s. The electrons are constrained to move along helical paths, driven by a rotating electric field in which the entire circuit is immersed. The electric field remains roughly orthogonal to the major axis of the helix and confines each charge carrier to a fraction of a turn of a single helical loop, moving it like water in an Archimedean screw. Each loop could in principle hold an independent carrier, permitting high information density. One computationally universal logic operation involves two helices, one of which splits into two "descendant" helices.

At the point of divergence, differences in the electrostatic potential resulting from the presence or absence of a carrier in the adjacent helix controls the direction taken by a carrier in the splitting helix. The reverse of this sequence can be used to merge two initially distinct helical paths into a single outgoing helical path without forcing a dissipative transition. Because these operations are both logically and thermodynamically reversible, energy dissipation can be reduced to extremely low levels. ... It is important to note that this proposal permits a single electron to switch another single electron, and does not require that many electrons be used to switch one electron. The energy dissipated per logic operation can likely be reduced to less than 10^{-27} joules at a temperature of 1 Kelvin and a speed of 10 gigahertz, though further analysis is required to confirm this. Irreversible operations, when required, can be easily implemented and should have a dissipation approaching the fundamental limit of $\ln 2 \times kT$. One study not conducted by Ames or partners is particularly worth mentioning since it places a loose lower bound on the computational capabilities of molecular nanotechnology. Drexler designed a number of computer components using small diamondoid rods with knobs that allow or prevent movement to accomplish computation. While this tiny mechanical Babbage Machine is probably not an optimal computational engine, its calculated performance for a desktop computer is 10^{18} MIPS — about a million times more powerful than the largest supercomputer that exists today. Note that with very fast computation energy use and heat dissipation become a severe problem. One approach to addressing this issue is reversible logic.

Drexler proposed a nanotechnology based on diamond and investigated its potential properties. In particular, he examined applications for materials with a strength similar to that of diamond (69 times strength/mass of titanium). This would require a very mature nanotechnology constructing systems by placing atoms on diamond surfaces one or a few at a time in parallel. Assuming diamondoid materials, McKendree predicted the performance of several existing single-stage-to-orbit (SSTO) vehicle designs. The predicted payload to dry mass ratio for these vehicles using titanium as a structural material varied from < 0 (the vehicle won't work) to 36 per cent, i.e., the vehicle weighs substantially more than the payload. With hypothetical diamondoid materials the ratios varied from 243 per cent to 653 per cent, i.e., the payload weighs far more than the vehicle. Using a very simple cost model ($1000 per vehicle kilogram) sometimes used in the aerospace industry, he estimated the cost per kilogram launched to low-Earth-orbit for diamondoid structured vehicles should be $153-412. This would meet NASA's 2020 launch to orbit cost goals. Estimated costs for titanium structured vehicles varied from $16,000-59,000/kg. Although this cost model is probably adequate for comparison, the absolute costs are suspect.

Drexler used a more speculative methodology to estimate that a four passenger SSTO weighing three tons including fuel could be built using a mature nanotechnology. Using McKendree's cost model, such a vehicle would cost about $60,000 to purchase — the cost of today's high-end luxury automobiles. These studies assumed a fairly advanced nanotechnology capable of building diamondoid materials. In the nearer term, it may be possible to develop excellent structural materials using carbon nanotubes. Carbon nanotubes have a Young's modulus of approximately one terapascal— comparable to diamond. Issacs and Pearson proposed a space elevator—a cable extending from the Earth's surface into space with a center of mass at geosynchronous altitude. If such a system could be built, it should be mechanically stable and vehicles could ascend and descend along the cable at almost any reasonable speed using electric power (actually generating power on the way down). The first incredibly difficult problem with building a space elevator is strength of materials. Maximum stress is at geosynchronous altitude so the cable must be thickest there and taper exponentially as it approaches Earth. Any potential material may be characterized by the taper factor—the ratio between the cable's radius at geosynchronous altitude and at the Earth's surface. For steel the taper factor is tens of thousands—clearly impossible. For diamond, the taper factor is 21.9 including a safety factor.

Diamond is, however, brittle. Carbon nanotubes have a strength in tension similar to diamond, but bundles of these nanometer-scale radius tubes shouldn't propagate cracks nearly as well as the diamond tetrahedral lattice. Thus, if the considerable problems of developing a molecular nanotechnology capable of making nearly perfect carbon nanotube systems approximately 70,000 kilometers long can be overcome, the first serious problem of a transportation system capable of truly large scale transfers of mass to orbit can be solved. The next immense problem with space elevators is safety—how to avoid dropping thousands of kilometers of cable on Earth if the cable breaks. Active materials may help by monitoring and repairing small flaws in the cable and/or detecting a major failure and disassembling the cable into small elements. Drexler calculates that lightsails made of 20 nm aluminum in tension should achieve an outward acceleration of ~14 km/s per day at Earth orbit with no payload and minimal structural overhead. For comparison, the delta V from low Earth to geosynchronous orbit is 3.8 km/s. Lightsails generate thrust by reflecting sunlight. Tension is achieved by rotating the sail. The direction of thrust is normal to the sail and away from the Sun. By directing thrust along or against the velocity vector, orbits can be lowered or raised. This form of transportation requires no reaction mass and generates thrust continuously, although the instantaneous acceleration is small so sails cannot operate in an atmosphere and must be large for even moderate payloads. Today, the smallest feature size in production systems is about 250 nanometers — the smallest feature size in computer chips. Since atoms are an angstrom or so across and carbon nanotubes have a diameter as small as 0.7 nanometers, atomically precise molecular machines can be smaller than current MEMS devices by two to three orders of magnitude in each dimension, or six to nine orders of magnitude smaller in volume (and mass). For example, the size of the kinesin motor, which transports material in cells, is 12 nm. Han computationally demonstrated that molecular gears fashioned from single-walled carbon nanotubes with benzyne teeth should operate well at 50-100 gigahertz. These gears are about two nanometers across. Han computationally demonstrated cooling the gears with an inert atmosphere. Srivastava simulated powering the gears using alternating electric fields generated by a single simulated laser. In this case, charges were added to opposite sides of the tube to form a dipole. To make active materials, a material might be filled with nano-scale sensors, computers, and actuators so the material can probe its environment, compute a response, and act. Although this document is concerned with relatively simple artificial systems, living tissue may be thought of as an active material. Living tissue is filled with protein machines which gives living tissue properties (adaptability, growth, self-repair, etc.) unimaginable in conventional materials. Active materials can theoretically be made entirely of machines. These are sometimes called swarms since they consist of large numbers of identical simple machines that grasp and release each other and exchange power and information to achieve complex goals. Swarms change shape and exert force on their environment under software control.

Although some physical prototypes have been built, at least one patent issued, and many simulations run, swarm potential capabilities are not well analyzed or understood. We briefly discuss some concepts here. Michael proposes brick-shaped machines of various sizes that slide past each other to assume a variety of shapes. He has generated a large number of videos showing computer simulations of simple motions. Although his web site contains rather extravagant claims, this work has received a U.K. patent. Yim built a small swarm with macroscopic (size in inches) components called polypod, built a simulator of polypod, and programmed it to move in various ways to study locomotion. There are two brick shaped components in polypod, one of which has two prismatic joints linked by a revolute joint. The second component is a cubic connector with no mechanical motion. Polypod is programmed by tables for each member of the swarm. Each member is programmed to move at various speeds in each degree of freedom for certain amounts of time. The swarm components are implicitly synchronized so there is no clock signal. Hall proposes a swarm with 10 micron dodecahedral components each with 12 arms that can move in and out, rotate a little, and grab and release each other. This concept is called the "utility fog." Hall estimates that the utility fog would have a density of 0.2, tensile strength of 1000 psi in action and 100,000 psi in a passive mode, and have a maximum shear rate of 100 km/second/meter.

Bishop proposes a swarm consisting of 100 nanometer brick-shaped components that slide past each other to change shape. Globus proposes a swarm with two kinds of components — edges and nodes. The terms "node" and "edge" are chosen to correspond to those in graph theory. The roughly spherical nodes are capable of attaching to five edges (for a tetrahedral geometry with one free edge per node) and rotating each edge in pitch and yaw. The rod-like edges are capable of changing length, rotating around their long axis, and attaching/detaching to/from nodes. Component design, power distribution and control software are significant challenges for swarm development. Consider that with 10 micron components a cubic meter of swarm would contain about 10^{15} devices, each with an internal computer communicating with its neighbors to accomplish a global task. NASA's mission is divided into five enterprises: Mission to Planet Earth, Aeronautics, Human Exploration and Development of Space, Space Science, and Space Technology. We will examine some potential nanotechnology applications in each area. The Earth Observing System (EOS) will use satellites and other systems to gather data on the Earth's environment. The EOS data system will need to process and archive >terabyte per day for the indefinite future. Simply storing this quantity of data is a significant challenge—each day's data would fill about 1,000 DVD disks. With projected nanocomputer processing speeds of 10^{18} MIPS, a million calculations on each byte of one day's data would take one second on the desktop. Given a mature nanotechnology, it should be possible to build sensors in balloon-borne systems approximately the size of bacteria. With replication based manufacturing, these should be quite inexpensive. If the serious communication and control problems can be solved, one can imagine spreading billions of tiny lighter-than-air vehicles into the atmosphere to measure wind currents and atmospheric composition. A similar approach might be taken in the oceans—note that the oceans are full of floating microscopic living organisms that can sense and react to their environment. Smart dust might sense the environment, note the location via a GPS-like system, and store that information until close enough to a data-collection point to transfer the data to the outside world. The strength of materials and computational capabilities previously discussed for space transportation should also allow much more advanced aircraft. Stronger, lighter materials can obviously make aircraft with greater lift and range. More powerful computers are invaluable in the design stage and of great utility in advanced avionics.

MEMS technology has been used to replace traditional large control structures on aircraft with large numbers of small MEMS controlled surfaces. This control system was used to operate a model airplane in a windtunnel. Nanotechnology should allow even finer control—finer control than exhibited by birds, some of which can hover in a light breeze with very little wing motion. Nanotechnology should also enable extremely small aircraft. A reasonably advanced nanotechnology should be able to make simple atomically precise materials under software control. If the control is at the atomic level, then the full range of shapes possible with a given material should be achievable. Aircraft construction requires complex shapes to accommodate aerodynamic requirements. With molecular nanotechnology, strong complex-shaped components might be manufactured by general purpose machines under software control. The aeronautics mission is responsible for launch vehicle development. Payload handling is an important function. Very efficient payload handling might be accomplished by a very advanced swarm. The sequence begins by placing each payload on a single large swarm located next to the shuttle orbiter. The swarm forms itself around the payloads and then moves them into the payload bay, arranging the payloads to optimize the center of gravity and other considerations. The swarm holds the payload in place during launch and may even damp out some launch vibrations. On orbit, satellites can be launched from the payload bay by having the swarm give them a gentle push. The swarm can then be left in orbit, perhaps at a space station, and used for orbital operations. This scenario requires a very advanced swarm that can operate in an atmosphere and on orbit in a vacuum. Besides the many and obvious difficulties of developing a swarm for a single environment, this provides additional challenges. Note that a simpler swarm might be used for aircraft payload handling. Aerospace vehicles often require complex checkout procedures to insure safety and reliability. This is particularly true of reusable launch vehicles. A very advanced swarm with some special purpose appendages might be placed on a vehicle. It might then spread out over the vehicle and into all crevices to examine the state of the vehicle in great detail.

Nanotechnology-enabled Earth-to-orbit transportation has the greatest potential to revolutionize human access to space by dropping the current $10,000 per pound cost of launch, but this was discussed above. Other less dramatic technologies include: Space structures with a long design life (such as space station modules) need high-reliability materials that do not degrade. Active materials might help. The machines monitor structural integrity at the sub-micrometer scale. When a portion of the material becomes defective, it could be disassembled and then correctly reassembled. It should be noted that bone works somewhat along these lines. It is constantly being removed and added by specialized cells. To effect timely repairs, space stations require a large store of spare parts and tools that are rarely used. A mature nanotechnology might create a "matter compiler," a machine that converts raw materials into a wide variety of products under software control. Contemporary examples of very limited matter compilers are numerically controlled machines and polypeptide sequencers. With a substantially more capable nanotechnology-based matter compiler, a space station crew could simply make spare parts and tools as needed. The programs could be stored on-board or on the ground. New tools invented on Earth could be transferred as software to the station for manufacture. Once used, unneeded tools and broken parts could be ionized in a solar furnace, transferred using controlled magnetic fields, and the constituent atoms stored for later manufacture into new products. An advanced nanotechnology might be able to build filters that dynamically modify themselves to attract the contaminant molecules detected by the air and water quality sensors. Once attached to the filter, the filter could in principle move the offending molecules to a molecular laboratory for modifications to useful or at least inert products. A swarm might implement such an active filter if it was able to dynamically manufacture proteins that could bind contaminant molecules. The protein and bound contaminant might then be manipulated by the swarm for transportation.

With a sufficiently advanced nanotechnology it might even be possible to directly generate food by non-biological means. Then agriculture waste in a self-sufficient space colony could be converted directly to useful nutrition. Making this food attractive will be a major challenge. Sleeping crew members in the shuttle experience considerable pain and sleep disruption when the reaction control system fires and they collide with the cabin walls. If crew members were connected to the walls by a swarm, the swarm could absorb most or all of the force before the crew member struck the wall. The swarm could then gradually return the crew member to center (without the oscillations associated with bungee cords) in preparation for the next firing. For resupply, spacecraft docking is a frequent necessity in space station operations. When two spacecraft are within a few meters of each other, a swarm could extend from each, meet in the middle, and form a stable connection before gradually drawing the spacecraft together. A swarm could support space-suited astronauts in simulated partial-g environments by holding them up appropriately. The swarm moves in response to the astronaut's motion providing the appropriate simulation of partial or 0 gravities. Tools and other objects are also manipulated by the swarm to simulate non-standard gravity. Active nanotechnology materials (see active materials) might enable construction of a skin-tight space suit covering the entire body except the head (which is in a more conventional helmet). The material senses the astronaut's motions and changes shape to accommodate it. This should eliminate or substantially reduce the limitations current systems place on astronaut range of motion. In situ resource utilization is undoubtedly necessary for large scale colonization of the solar system. Asteroids are particularly promising for orbital use since many are in near Earth orbits. Moving asteroids into low Earth orbit for utilization poses a safety problem should the asteroid get out of control and enter the atmosphere. Very small asteroids can cause significant destruction. The 1908 Tunguska explosion, which was calculated to be a 60 meter diameter stony asteroid, leveled 2,200 km^2 of forest. Hills calculated that 4 meter diameter iron asteroids are near the threshold for ground damage. Both these calculations assumed high collision speeds. At a density of 7.7 g/cm^3 Babadzhanov, a 3 meter diameter asteroid should have a mass of about 110 tons. Rabinowitz estimates that there are about one billion ten meter diameter near Earth asteroids and there should be far more smaller objects. For colonization applications one would ideally provide the same radiation protection available on Earth. Each square meter on Earth is protected by about 10 tons of atmosphere. Therefore, structures orbiting below the van Allen belts would like 10 tons/meter^2 surface area shielding mass. This would dominate the mass requirements of any system and require one small asteroid for each 11 meter^2 of colony exterior surface area. A 10,000 person cylindrical space colony such as Lewis One with a diameter of almost 500 meters and a length of nearly 2000 meters would require a minimum of about 90,000 retrieval missions to provide the shielding mass.

The large number of missions required suggests that a fully automated, replicating nanotechnology may be essential to build large low Earth orbit colonies from small asteroids. A nanotechnology swarm along with an atomically precise lightsail is a promising small asteroid retrieval system. Lightsail propulsion insures that no mass will be lost as reaction mass. The swarm can control the lightsail by shifting mass. When a target asteroid is found, the swarm spreads out over the surface to form a bag. The interface to the sail must be active to account for the rotation of the asteroid—which is unlikely to have an axis-of-rotation in the proper direction to apply thrust for the return to Earth orbit. The active interface is simply swarm elements that transfer between each other to allow the sail to stay in the proper orientation. Of course, there are many other possibilities for nanotechnology based retrieval vehicles. Extraterrestrial materials brought into orbit could be fed into a high-temperature solar furnace and partially ionized. Magnetic fields might then be used to separate the nuclei. These are fed in appropriate quantities to a matter-compiler to build the products desired. Several authors, including Freitas have speculated that a sufficiently advanced nanotechnology could examine and repair cells at the molecular level. Should this capability become available—presumably driven by terrestrial applications—the small size and advanced capabilities of such systems could be of great utility on long duration space flights and on self-sufficient colonies.Self-replicating systems permit efforts of great scope to be pursued economically. Adjusting the environment on another planet to suit the tastes of humans is one such undertaking. Heating and cooling can be achieved by (among many other methods) using space-based mirrors.

Chemical modifications of the planetary surface and atmosphere can be achieved in relatively short periods by the use of self-replicating systems that absorb sunlight and raw materials, and convert them into the desired products. Much as plants changed the environment of the earth to what we see today, so self-replicating molecular manufacturing systems might more rapidly convert the environments of other planets. As interstellar trips might last many years, the ability to conserve supplies by maintaining some crew members in a suspended state would be useful. An extremely advanced nanotechnology might use molecular manipulations of each cell to provide (a) better methods of slowing or suspending the metabolic activity of crew members and (b) better methods of restoring metabolic activity to a normal state when the destination is reached. Molecular manufacturing should enable the creation of very precise mirrors. Unlike lightsail applications, telescope mirrors require a very precise and somewhat complicated shape. A swarm with special purpose appendages capable of bonding to the mirror might be able to achieve and maintain the desired shape. A very advanced nanotechnology would be capable of imaging and then removing the surface atoms of an extra-terrestrial sample. By removing successive surface layers the location of each atom in the sample might be recorded, destroying the sample in the process.

This data could then be sent to Earth. Besides requiring a very advanced nanotechnology, there is a more fundamental—but not necessarily fatal—problem: as the outside layer of atoms is removed the next layer may rearrange itself so the sample is not necessarily perfectly recorded. Solar Power—Low Earth orbit spacecraft generally depend on solar cells and batteries for power. For energy collection, molecular manufacturing can be used to make solar photovoltaic cells at least as efficient as those made in the laboratory today.

Efficiencies can therefore be > 30 per cent. In space applications, a reflective optical concentrator need consist of little more than a curved aluminum shell < 100 nanometers thick (photovoltaic cells operate with higher efficiency at high optical power densities). A metal fin with a thickness of 100 nanometers and a conduction path length of 100 microns can radiate thermal energy at a power density as high as 1000 W/m^2 with a temperature differential from base to tip of < 1 K.

Accordingly, solar collectors can consist of arrays of photovoltaic cells several microns in thickness and diameter, each at the focus of a mirror of ~100 micron diameter, the back surface of which serves as a ~100 micron diameter radiator. If the mean thickness of this system is ~1 micron, the mass is ~10^{-3} kg/m^2 and the power per unit mass, at Earth's distance from the Sun, where the solar constant is ~1.4 kW/m^2, is > 10^5 W/kg."

By comparison, the U.S. built Photovoltaic Panel Module solar cells currently used on the Mir Space Station and planned for use on the International Space Station generate about 118 W/kg. Fuel Cells—A critical component in hydrogen/oxygen fuel cells is the PEM (Proton Exchange Membrane). This membrane must (a) permit the passage of protons while (b) blocking everything else. Present membranes do a rather poor job. One group at Ames is designing and computationally testing PEMs to study possible energy mechanisms in early life. While these studies are not meant to design optimal membranes for fuel cell use, the basic knowledge and approach may be of value. Another proposal is to design a diamond membrane a few nanometers thick with "proton pores."

The pores might be lined with fluorine, oxygen and nitrogen to create a region with a high proton affinity. In addition, a positionally controlled platinum might be held at the mouth of the pore to verify that H_2 can be catalytically split into H+ and e–, and that the barrier for migration of the H+ into the pore is modest in size. Nanotechnology must provide precise control over the manufacturing process of the diamondoid PEM since the pores must be made very precisely. Studies of H_2 absorption and packing in carbon nanotubes and nanoropes are in progress at NASA Ames and elsewhere. Nanotubes provide large pore sizes and nanoropes have different pore sizes depending on interstitial and other locations. Dillon estimated that the single walled nanotubes in their sample contained 5 to 10 per cent by weight of H_2. The nanotubes were about 0.1 to 0.2 per cent by weight of the total sample. Computational studies at Ames suggest that to store 7-10 per cent H_2 in single walled nanotubes at room temperature the H_2s must be stored inside the tubes, not merely adsorbed on the walls. This work suggests that carbon nanotubes might be developed into an excellent H_2 storage medium within 3-5 years. Calculations with oxygen suggest that a diamondoid sphere ~0.1 microns in diameter should easily hold oxygen at ~1,000 atmospheres. While higher pressures are feasible, they offer declining returns. At higher pressures, the pressure-volume relationship becomes severely non-linear and the density approaches a limiting value. Other gases might also be stored if diamondoid spheres can be built, but the analysis has not been done. High strength light-weight materials will allow greater efficiency of energy storage as angular momentum. Many kinds of ultraminiature electromechanical devices have utility on a miniaturized space craft. It has been shown that manipulating carbon nanotubes changes their electrical properties. This might be exploited to build nanometer scale strain devices.

This may be achievable within 3-5 years, and simulations along these lines are in progress. Similar results have been achieved experimentally with C_{60}. The electrical properties of a C_{60} molecule were changed by applying pressure to the molecule with an SPM tip. Smaller, lighter spacecraft are cheaper to launch (current costs are about $10,000/lb) and generally cheaper to build. Diamondoid structural materials can radically reduce structural mass, miniaturized electronics can shrink the avionics and reduce power consumption, and atomically precise materials and components should shrink most other subsystems. Thermal protection is crucial for atmospheric reentry and other tasks.

The carbon nanotubes under investigation at NASA Ames and elsewhere may play a significant role. Most production processes for carbon nanotubes create a tangled mat of nanotubes that has a very low mass-to-volume ratio.

Like graphite, the tubes should withstand high temperatures but the tangled mat should prevent them from ablating. This may lead to high temperature applications. Many of the applications discussed here are speculative to say the least.

However, they do not appear to violate the laws of physics. Something similar to these applications at these performance levels should be feasible if we can gain complete control of the three-dimensional structure of materials, processes and devices at the atomic scale. How to gain such control is a major, unresolved issue. However, it is clear that computation will play a major role regardless of which approach—positional control with replication, self-assembly, or some other means—is ultimately successful. Computation has already played a major role in many advances in chemistry, SPM manipulation, and biochemistry.

As we design and fabricate more complex atomically precise structures, modeling and computer aided design will inevitably play a critical role. Not only is computation critical to all paths to nanotechnology, but for the most part the same or similar computational chemistry software and expertise supports all roads to molecular nanotechnology. Thus, even if NASA's computational molecular nanotechnology efforts should pursue an unproductive path, the expertise and capabilities can be quickly refocused on more promising avenues as they become apparent.

As nanotechnology progresses we may expect applications to become feasible at a slowly increasing rate. However, if and when a general purpose programmable assembler/ replicator can be built and operated, we may expect an explosion of applications. From this point, building new devices will become a matter of developing the software to instruct the assembler/replicators. Development of a practical swarm is another potential turning point. Once an operational swarm that can grow and divide has been built, a large number of applications become software projects. It is also important to note that the software for swarms and assembler/replicators can be developed using simulators — even before operational devices are available.

Nanotechnology advocates and detractors are often preoccupied with the question "When?" There are three interrelated answers to this question:

(i) Nobody knows. There are far too many variables and unknowns. Beware of those who have excessive confidence in any date.

(ii) The time-to-nanotechnology will be measured in decades, not years. While a few applications will become feasible in the next few years, programmable assembler/replicators and swarms will be extremely difficult to develop.

(iii) The time-to-nanotechnology is very sensitive to the level of effort expended. Resources allocated to developing nanotechnology are likely to be richly rewarded, particularly in the long term.

Genetic Engineering

Genetic engineering (GE) is the manipulation of genetic material (ie, DNA or genes) in a cell or an organism in order to produce desired characteristics and to eliminate unwanted ones. GE includes a range of different techniques with many different uses, and can be applied to plants, animals and humans. For example, the genetic modification of food is a form of GE that involves manipulating the cells of plants such as maize, to increase the yields, make it more nutritious and to make it drought- and disease-resistant. However, the most contentious type of GE is definitely related to its applications in humans. GE in humans has opened up a Pandora's box of possibilities as it can be used for both the miraculous and the sinister.

The cloning of Dolly the sheep in 1996 was a very important event. Until then, the cloning of a human was only possible in theory. Recent films and TV programmes such as *Gattaca, Mutant X, Dark Angel* and books such as *Brave New World* all focus on possible consequences of this technology – but what is the real deal? If you were told that there was a clone of you sitting in the next room, what would you expect the clone to look and act like? Probably, exactly the same as you. But, despite all the movies and TV programmes that have explored the possibility of exact clones, it is highly unlikely that a clone of you would look exactly the same and would certainly not act exactly the same.

A clone is not completely genetically identical, as there are small differences in the genetic make-up just as there are with identical twins. Despite the fact that identical twins come from the same egg, after a while one begins to notice the differences between them in order to tell them apart. It has been discovered – by doing a number of studies on identical twins brought up in the same environment – that genes mysteriously react differently to the same environment. While cloning a whole human is certainly the ultimate challenge for genetic engineers, cloning is not limited to this goal. Cloning can be, and is, done on a much smaller scale and could involve no more than just the cloning of a single cell. Therefore, cloning can be divided into two types: (i) reproductive cloning (which is the cloning of a whole organism); and (ii) therapeutic cloning (which is the cloning of cells or even organs or other tissue for transplant purposes). Due to the fact that genetic differences are likely to exist between a clone and its donor, this uncertainty has led to many countries banning the reproductive cloning of humans. In order to make a clone of someone, one needs a living cell and a human egg (ovum).

The nucleus of the egg, which contains the DNA, is removed and replaced with the nucleus from the cell of the person/animal to be cloned. A short electrical pulse then stimulates the egg to start dividing and the embryo is then implanted into the womb where it develops into a duplicate of the person that donated the cell nucleus. Clones created in this way are not 100% genetically identical, as there is some DNA from the original egg cell that is found outside the nucleus (mitochondrial DNA). In therapeutic cloning an embryo is created in the same way as reproductive cloning, but it is not implanted into the womb of a woman.

Instead, *stem cells* are extracted after the embryo starts dividing in the first 14 days after fertilisation, which kills the embryo. Stem cells are special cells with the ability to reproduce and become one of 300 types of cells, eg, skin, liver cell, hair or blood cells. These cells are then used to grow the specific type of tissue or organ that is needed and has the advantage of being genetically identical to the patient who donated it, eliminating the problem of organ or tissue rejection. Currently, if someone has an organ transplant, there is quite a high possibility that their body will reject the foreign organ and so they not necessary have to suppress the immune system to lessen the chances of this happening. Stem cells could potentially be used to repair damaged or defective tissues around the body, such as the cells in the pancreas that stop producing insulin in diabetics. The problem with embryonic stem cells is that many people feel that by using a human embryo and then killing it, you are actually killing a potential person.

It is for this reason that the United States and other

countries are calling for a global ban on all human cloning, including the use of embryonic stem cells. As a result, there has been new research into the potential of what is known as "adult stem cells". More than 30 years ago it was discovered that, in addition to being found in embryos, stem cells are also found in adults. Adult stem cells can be removed from a person without causing any harm. Until very recently, it was thought that adult stem cells were only suitable for cloning a few types of cells and tissues.

However, new studies have found adult stem cells with almost the same abilities as embryonic stem cells in various human tissues, including: the spinal cord, the brain, connective tissue and in the blood of the umbilical cord. These studies have shown that it is possible for us to manipulate cells to take on new functions/start doing different jobs simply by placing them in different environments and "reprogramming" them. For example, neural (brain) stem cells in mice have been transformed into the blood stem cells that produce the different types of blood cells. In other animals, bone marrow stem cells have become brain cells and liver cells.

In humans, bone marrow stems cells of a patient have been transplanted into the areas of the heart damaged by a heart attack, triggering new growth of heart tissue. Although adult stem cells are already being used successfully to treat a number of diseases, there are a number of problems to be overcome, such as finding and extracting the adult stem cells.

Embryonic stem cells have not yet been used for therapy and research is still in the initial stages. Eventually, it may be that, rather than just one type, both types of stem cells will be used for different purposes – whichever proves to be the best for the situation. In the meantime, parallel research continues on both cell types. Although currently the risks of cloning outweigh the possible benefits, there are many different potential uses of human cloning technology:

(i) *Replacing organs and other tissues* – such as new skin for burn victims, brains cells for those with brain damage, spinal rod cells for the paralysed and complete new organs (hearts, liver, kidney and lungs). Pigs are also being genetically modified to make their organs more compatible with humans by removing the gene that causes rejection. People could have their appearance changed (cosmetic surgery) using their own cloned tissue and accident victims and amputees could also benefit from this tissue regeneration.

(ii) *Infertility* – human cloning provides couples and individuals who are unable to have children with another potential option.

(iii) *Replacement of a lost child* – parents who have lost a child through an accident or an illness could clone an identical "replacement" child.

(iv) *Creating "donor" people* – cloned people could be created to provide a source of transplant material.

(v) *Gene therapy* – cloning technology could be used to prevent, treat and cure genetic disorders by changing the expression of a person's genes. This technology may also provide the cure for cancer by revealing how cells are switched on and off. Gene therapy could be used to treat somatic (body) cells where the change is not passed on to children, or germ (egg and sperm) cells where the changes are passed on.

(vi) *Saving endangered species* – by boosting their numbers through creating clones. However, since clones are almost genetically identical, the genetic diversity of the species would not be increased.

(vii) *Reversal of the ageing process* – once more is understood about the role that our genes play in the ageing process. However, some of the above uses carry with them some serious ethical implications.

Dolly the sheep was created in 1996 using the cloning methods outlined above. Although Dolly was born looking normal, she has suffered from several problems associated with the cloning technique, including premature arthritis, which is thought to be a side-effect of the cloning. Other problems with the current cloning techniques, include:

Allowing Cloning

Around the world, different countries have different rules relating to cloning – some don't allow reproductive cloning, but do allow therapeutic cloning, while other countries allow both types. In South Africa currently, the Human Tissue Act prohibits both therapeutic and reproductive cloning of humans. This situation will change when chapters 6 and 8 of the National Health Act are promulgated - and therapeutic cloning will then be allowed under strict conditions, requiring Ministerial permission, but reproductive cloning on humans will remain strictly prohibited. In China, there are no laws against cloning either. However, with the huge population problem and the policy of only one child per family, there is no interest in reproductive cloning. On the other hand, there is huge interest in therapeutic cloning and, as a result, there is a lot of it on the go in China.

Some Chinese cultures believe that humans only become people when they participate in society – so, according to these cultures, embryos and foetuses are not considered to be human beings, thereby eliminating any ethical problems surrounding the creation and destruction of embryos to get the required stem cells. The Chinese government is funding extensive research and drawing back many Chinese scientists from overseas, to undertake work they would not be allowed to do elsewhere. With easy access and no limits on obtaining the embryonic material they require, it is expected that Chinese scientists will race ahead of the rest of the world in therapeutic cloning technologies.

In the UK there are clear rules banning reproductive cloning, but scientists in both the UK and Israel are allowed to generate new embryonic cell lines for therapeutic research. The law in Germany bans the extraction of stem cells from human embryos for research within the country, but in 2002 a new law was passed allowing some human embryonic stem cells to be imported. This means that German scientists are allowed to undertake work on embryonic stem cells if they originate from outside Germany.

In the United States there is no public funding for research on embryonic stem cells. Therefore, although there is no law against therapeutic cloning in the US, there is almost no public research happening due to the lack of money and access to the embryonic material. Simultaneously, an announcement was recently made stating that US public funds (about US$1,4 million, about R11,2 million) will be provided for studies using adult stem cells instead of embryonic stem cells. These different rules for different countries mean that if a scientist is banned from undertaking cloning in his/her home country, he/she can simply move to a country where it is allowed. To prevent this from happening, the US and about 30 other countries want a global ban on all forms of cloning. However, many countries don't agree and thus no progress is being made on implementing this ban. Genetic engineering, genetic modification (GM), and the now-deprecated gene splicing are terms for the process of manipulating genes, usually outside the organism's normal reproductive process. It often involves the isolation, manipulation and reintroduction of DNA into cells or model organisms, usually to express a protein.

The aim is to introduce new characteristics such as making a crop resistant to an herbicide, introducing a novel trait, or producing a new protein or enzyme. Examples include the production of human insulin through the use of modified bacteria, the production of erythropoietin in Chinese Hamster Ovary cells, and the production of new types of experimental mice such as the OncoMouse (cancer mouse) for research, through genetic redesign. Since a protein is specified by a segment of DNA called a gene, future versions of that protein can be modified by changing the gene's underlying DNA.

One way to do this is to isolate the piece of DNA containing the gene, precisely cut the gene out, and then reintroduce (splice) the gene into a different DNA segment. Daniel Nathans and Hamilton Smith received the 1978 Nobel Prize in physiology or medicine for their isolation of restriction endonucleases, which are able to cut DNA at specific sites. Together with ligase, which can join fragments of DNA together, restriction enzymes formed the initial basis of recombinant DNA technology.

Genetic engineering, the use of various methods to manipulate the DNA (genetic material) of cells to change hereditary traits or produce biological products. The techniques include the use of hybridomas (hybrids of rapidly multiplying cancer cells and of cells that make a desired antibody) to make monoclonal antibodies monoclonal antibody, an antibody that is mass produced in the laboratory from a single clone and that recognizes only one antigen. Monoclonal antibodies are typically made by fusing a normally short-lived, antibody-producing B cell (see immunity) to a fast-growing cell, such as a cancer cell (sometimes referred to as an "immortal" cell).; gene splicing or recombinant DNA, in which the DNA of a desired gene is inserted into the DNA of a bacterium, which then reproduces itself, yielding more of the desired gene; and polymerase chain reaction polymerase chain reaction (pOl`-mYrâs') (PCR), laboratory process in which a particular DNA segment from a mixture of DNA chains is rapidly replicated, producing a large, readily analyzed sample of a piece of DNA; the process is sometimes called DNA amplification, which makes perfect copies of DNA fragments and is used in DNA fingerprinting DNA fingerprinting or DNA profiling, any of several similar techniques for analyzing and comparing DNA from separate sources, used especially in law enforcement to identify suspects from hair, blood, semen, or other biological materials found at the scene of a violent crime.

Monoclonal antibody, an antibody that is mass produced in the laboratory from a single clone and that recognizes only one antigen. Monoclonal antibodies are typically made by fusing a normally short-lived, antibody-producing B cell (see immunity immunity, ability of an organism to resist disease by identifying and destroying foreign substances or organisms. Although all animals have some immune capabilities, little is known about nonmammalian immunity. Mammals are protected by a variety of preventive mechanisms, some of them nonspecific (e.g., barriers, such as the skin), others highly specific (e.g., the response of antibodies). Nonspecific Defenses) to a fast-growing cell, such as a cancer cell (sometimes referred to as an "immortal" cell). The resulting hybrid cell, or hybridoma, multiplies rapidly, creating a clone that produces large quantities of the antibody.

Monoclonal antibodies engendered much excitement in the medical world and in the financial world in the 1980s, especially as potential cures for cancer. They have been used in laboratory research and in medical tests since the mid-1970s, but their effectiveness in disease treatment has been limited. By the mid-1990s, however, some of the technical problems had been overcome. Experimental cancer therapies have used drugs, radioactive materials, or immune killer cells attached to monoclonal antibodies that, when injected into patients, home in on antigens that grow only on the surface of cancer cells.

Genetically engineered products include bacteria designed to break down oil slicks and industrial waste products, drugs (human and bovine growth hormones, human insulin, interferon), and plants that are resistant to diseases, insects, and herbicides, that yield fruits or vegetables with desired qualities, or that produce toxins that act as pesticides. Genetic engineering techniques have also been used in the direct genetic alteration of livestock and laboratory animals (see pharming pharming (fär'm-ng), the use of genetically altered livestock, such as cows, goats, pigs, and chickens, to produce medically useful products. In pharming, researchers first create hybrid genes using animal DNA and the human or other gene that makes a desired substance, such as a hormone.). Genetically engineered products usually require the approval of at least one U.S. government agency, such as the Dept. of Agriculture, the Food and Drug Administration, or the Environmental Protection Agency.

Because genetic engineering involves techniques used to obtain patents on human genes and to create patentable living organisms, it has raised many legal and ethical issues. The safety of releasing into the environment genetically altered organisms that might disrupt ecosystems has also been questioned. The discovery in 2001 of genetically engineered DNA in native Mexican corn varieties made concerns of genetic pollution actual, and led some scientists to worry that the spread of transgenes through cross-pollination could lead to a reduction in genetic diversity in important crops.

Imports of genetically modified corn, soybeans, and other crops have been curtailed or limited in some countries, and the vast majority of such crops are grown in just a handful of nations. The Cartagena Protocol on Biosafety, which has been signed by more than 100 nations and took effect in Sept., 2003, requires detailed information on whether and how imported seeds, plants, animals, other organisms, and the like are genetically modified and permits a nation to bar those imports. The United States, however, is not party to the treaty. When Hungarian engineer Karl Ereky coined the word biotechnology in 1919 to describe the production of goods from raw materials with the aid of living organisms, he was describing techniques that had been employed by human beings for centuries. This includes techniques such as selective breeding, fermentation, hybridization, phyto-pharmacology, vaccination or the use of biomass for energy production.

However, the term biotechnology has evolved to include modern techniques with new applications, to what is now known as modern biotechnology. Modern biotechnology encompasses techniques such as genetic engineering (which is the direct intervention on the genetic makeup of an organisms usually by introducing foreign DNA into its gene pool by means that would not occur naturally), tissue culture and others. These modern techniques also take considerably less time to achieve the desirable changes in living organisms and carry greater precision than traditional techniques. In short, genetic engineering is a specific subset as a part of the more general subject of biotechnology.

Human genetic engineering refers to the the controlled

modification of the human genome. DNA provides the the genetic blueprint for all living organisms and can influence individuals' actions and abilities. With the advent of DNA research and the ability to change gene expressions, it is now possible that scientists may be able to change human capacities, whether they be physical, cognitive, or emotional. Human genetic engineering is still in its infancy, however, with current research generally restricted to animals or gene therapy. Healthy humans do not need gene therapy to survive, though it may prove helpful to treat certain diseases. Special gene modification research has been carried out on groups such as the 'bubble children' - those who's immune system do not protect them from the bacteria and irritants all around them. The first clinical trial of human gene therapy began in 1990, but (as of 2006) gene therapy is still experimental. Other forms of human genetic engineering are still theoretical, or restricted to fiction stories. Recombinant DNA research is usually performed to study gene expression and various human diseases. Some drastic demonstrations of gene modification have been made with mice and other animals, however: testing on humans is generally considered off-limits. In some instances changes are usually brought about by removing genetic material from one organism, and transferring them into another species.

This method is known as recombinant genetics. There are two main types of genetic engineering. Somatic modifications involve adding genes to cells other than egg or sperm cells. For example. if a person had a disease caused by a defective gene, a healthy gene could be added to the affected cells to treat the disorder. The distinguishing characteristic of somatic engineering is that it is non-inheritable, e.g. the new gene would not be passed to the recipient's offspring. Germline engineering would change genes in eggs, sperm, or very early embryos. This type of engineering is inheritable, meaning that the modified genes would appear not only in any children that resulted from the procedure, but in all succeeding generations. This application is by far the more consequential as it could open the door to the perpetual and irreversible alteration of the human species. There are two techniques researchers are currently experimenting:

- Viruses are good at injecting their DNA payload into human cells and reproducing it. By adding the desired DNA to the DNA of non-pathogenic virus (a pathogen is something that harms you. A non pathogenic virus is still a virus by definition, but will not cause you to get sick), a small amount of virus will reproduce the desired DNA and spread it all over the body.
- Manufacture large quantities of DNA, and somehow package it to induce the target cells to accept it, either as an addition to one of the original 46 chromosomes, or as an independent 47th human artificial chromosome.

Genetic engineering is most easily accomplished by making changes just after the egg and sperm have melded but before first division. In this way, the gene will be expressed throughout and will affect the recipients children, grandchildren, and all subsequent generations. This type of germline engineering is highly controversial and deemed inappropriate by most scientists. As of now, this is likely to take the form of gene therapy. This is less controversial as the engineered person can give consent, unlike germline engineering where subsequent generations are brought into the world with a genetic modification made to their body without their consent. Each human cell contains a different set of genes, so different genes could be transferred to different parts of the body. Such changes will not be hereditary unless the sex cells are engineered.

REFERENCES

Adachi S, et al. Cyclin A/cdk2 activation is involved in hypoxia-induced apoptosis in cardiomyocytes Circ Res. 88:408, 2001.

Agocha A, Lee H-W, Eghali-Webb M. Hypoxia regulates basal and induced DNA synthesis and collagen type I production in human cardiac fibroblasts: effects of TGF-beta, thyroid hormone, angiotensis II and basic fibroblast growth factor. J Mol Cell Cardiol 1997;29:2233-2244H.

Andres, V. and Walsh, K. (1996) Myogenin expression, cell cycle withdrawal and phenotypic differentiation are temporally separable events that precedes cell fusion upon myogenesis. J. Cell Biol., 132, 657-666.

Annual Report of the US Scientific Registry for Organ Transplantation and the Organ Procurement and Transplantation Network - 1990. Washingtion, DC: US Department of Health and Human Services: 1990.

Anversa P, and Nadal-Ginard B. Myocyte renewal and ventricular remodelling Nature 415, 240 - 243, 2002.

Aorta-associated CD34 hematopoietic cells in the early human embryo.

Asahara T, Takahashi T, Masuda H, Kalka C, Chen D, Iwaguro H, Inai Y, Silver M, Isner JM. VEGF contributes to postnatal neovascularization by mobilizing bone marrow-derived endothelial progenitor cells. EMBO J 1999; 18:3964- 3972.

Asahara, T. et al. Isolation of putative progenitor cells for endothelial angiogenesis. Science 1997; 275:964-967.

Beauchamp JR, Morgan JE, Pagel CN, Partridge TA (1999) Dynamics of myoblast transplantation reveal a discrete minority of precursors with stem celllike properties as the myogenic source. J Cell Biol 144:1113-1122.

Beltrami AP, et al. Evidence that human cardiac myocytes divide after myocardial infarction. *N Engl J Med* 344, 1750-7 (2001).

BZ Atkins, MT Hueman, JM Meuchel, MJ Cottman, KA Hutcheson, DA Taylor: Myogenic cell transplantation improves in vivo regional performance in infarcted rabbit myocardium. J Heart Lung Transpl 1999, 18: 1173-1180.

CE Murry, RW Wiseman, SM Schwartz, SD Hauschka: Skeletal myoblast transplantation for repair of myocardial necrosis. J Clin Invest 1996, 98: 2512-2523.

Choi K, Kennedy M, Kazarov A, Papadimitriou, Keller G. A common precursor for hematopoietic and endothelial cells. Development 1998;125:725-732.

Colucci WS. Molecular and cellular mechanisms of myocardial failure. Am J Cardiol 1997;80:15L-25L.

DA Taylor, BZ Atkins, P Hungspreugs, TR Jones, MC Reedy, KA Hutcheson, DD Glower, WE. Regenerating functional myocardium: improved performance after skeletal myoblast transplantation. Nature Medicine 1998, 4: 929-933.

Effects of enalapril on mortality in severe congestive heart failure. Results of the Cooperative North Scandinavian Enalapril Survival Study (CONSENSUS) Trial Study Group. N Engl J Med 1987; 316(23):1429-1435.

Elefanty AG, Robb L, Birner R, Begley CG. Hematopoietic-specific genes are not induced during in vitro differentiation of scl-null embryonic stem cells.

Folkman, J. Therapeutic angiogenesis in ischemic limbs. Circulation 1998; 97:108- 110.

Frazier OH, Rose EA, Macmanus Q, et al. Multicenter clinical evaluation of the Heartmate 1000 IP left ventricular assist device. Ann Thorac Surg 1992; 102:578-587.

Halevy, O., Novitch, B.G., Spicer, D.B., Skapek, S.X., Rhee, J., Hannon, G.J., Beach, D. and Lassar, A.B. (1995) Correlation of terminal cell cycle arrest of skeletal muscle with induction of p21 by MyoD. Science, 267, 1018- 1021.

Hescheler J, Fleischmann BK, Wartenberg M, Bloch W, Kolossov E, Ji G, Addicks K, Sauer H (1999) Establishment of ionic channels and signalling cascades in the embryonic stem cell-derived primitive endoderm and cardiovascular system. Cells Tissues Organs 165:153-164.

Heymans S, Luutun A, Nuyens D, et al. Inhibition of plasminogen activators or matrix metalloproteinases prevents cardiac rupture but impairs therapeutic angiogenesis and causes cardiac failure. Nat Med 1999;5:1135- 1142.

Hill MF, Singal PK. Right and left myocardial antioxidant responses during heart failure subsequent to myocardial infarction. Circulation 1997 96:2414-20.

Hochman JS and Choo H. Limitation of myocardial infarct expansion by reperfusion independent of myocardial salvage. Circulation 1987;75:299- 306.

Hodgetts SI, Beilharz MW, Scalzo T, Grounds MD (2000) Why do cultured transplanted myoblasts die in vivo? DNA quantification shows enhanced survival of donor male myoblasts in host mice depleted of CD4+ and CD8+ or NK1.1+ cells. Cell Transplant 9:489-502.

Hognes JR. In The Artificial Heart: Prototypes, Policies and Patients. Washingtion, DC: National Academy Press; 1991:1-312.

Jackson KA, Majka SM, Wang H, Pocius J, Hartley CJ, Majesky MW, Entman ML, Michael LH, Hirschi KK, Goodell MA (2001) Regeneration of ischemic cardiac muscle and vascular endothelium by adult stem cells. J Clin Invest 107:1395-1402.

Jaffredo T, Gautier R, Eichmann A, Dieterlen-Lievre F. Intraaortic hemopoietic cells are derived from endothelial cells during ontogeny. Development 1998;125:4575-4583.

Kajstura J, Leri A, Finato N, di Loreto N, Beltramo CA, Anversa P. Myocyte proliferation in end-stage cardiac failure in humans. Proc Natl Acad Sci USA 95, 8801-8805 (1998).

Kalka, C. et al. Transplantation of ex vivo expanded endothelial progenitor cells for therapeutic neovascularization. Proc. Natl. Acad. Sci. USA 2000; 97:3422-3427.

Kalkman EAJ, Bilgin YM, van Haren P, van Suylen R-J, Saxena PR, Schoemaker RG.Determinants of coronary reserve in rats subjected to coronary artery ligation or aortic banding. Cardiovasc Res 1996.

Kawano H, Do YS, Kawano Y, Starnes V, Barr M, Law RE, Hsueh WA. Angiotensin II has multiple profibrotic effects in human cardiac fibroblasts. Circulation 2000;101:1130-1137.

Kehat I, Kenyagin-Karsenti D, Snir M, Segev H, Amit M, Gepstein A, Livne E, Binah O, Itskovitz-Eldor J, Gepstein L. Human embryonic stem cells can differentiate into myocytes with structural and functional properties of cardiomyocytes. J Clin Invest. 2001; 108: 407-14.

Kellerman S, Moore JA, Zierhut W, Zimmer HG, Campbell J, Gerdes AM. Nuclear DNA content and nucleation patterns in rat cardiac myocytes from different models of cardiac hypertrophy. *J Mol Cell Cardiol* , 497-505 (1992).

Kennedy M, Firpo M, Choi K, Wall C, Robertson S, Kabrun N, Keller G. A common precursor for primitive erythropoiesis and definitive haematopoiesis.

Klug MG, Soonpaa MH, Koh GY, Field LJ (1996) Genetically selected cardiomyocytes from differentiating embryonic stem cells form stable intracardiac grafts. J Clin Invest 98:216-224.

Kocher A, Schuster M, Szabolcs M, Itescu S. Cardiomyocyte regeneration after neovascularization of ischemic myocardium by human bone marrowderived angioblasts. Nature Medicine 2002, in press.

Kocher AA, Schuster MD, Szabolcs MJ, Takuma S, Burkhoff D, Wang J, Homma S, Edwards NM, Itescu S. Neovascularization of ischemic myocardium by human bone-marrow-derived angioblasts prevents cardiomyocyte apoptosis, reduces remodeling and improves cardiac function. Nature Med.

L Field: Future therapy for cardiovascular disease. In Proceedings of the NHLBI Workshop Cell Transplantation: Future Therapy for Cardiovascular Disease? Columbia, MD; 1998.

Labastie M-C, Cortes F, Romeo P-H, Dulac C, Peault B. Molecular identity of hematopoietic precursor cells emerging in the human embryo. Blood 1998;92:3624-3635.

Lancet 347:279.

Levkau B, Koyama H, Raines EW, Clurman BE, Herren B, Orth K, Roberts JM, Ross R. Cleavage of p21cip1/waf1 and p27 kip1 mediates apoptosis in endothelial cells through activation of cdk2: role of a caspase cascade.

Li, Y., Jenkins, C.W., Nichols, M.A. and Xiong, Y. (1994) Cell cycle expression and p53 regulation of the cyclin-dependent kinase inhibitor p21. Oncogene, 9, 2261-2268 Steinman, R.A., Hoffman, B., Iro, A., Guillouf, C., Liebermann, D.A. and El-Houseini, M.E. (1994) Induction of p21 (WAF1/CIP1) during differentiation.

Liechty KW, MacKenzie TC, Shaaban AF, et al. Human mesenchymal stem cells engraft and demonstrate site-specific differentiation after in utero transplantation in sheep. Nat Med 2000, 6:1282-6.

Lin Y, Weisdorf DJ, Solovey A, Hebbel RP. Origins of circulating endothelial cells and endothelial outgrowth from blood. J Clin Invest 2000;105:71-77.

MacLellan WR and Schneider MD. Genetic dissection of cardiac growth control pathways Annu. Rev. Physiol. 2000. 62:289-320.

Mahon NG, O'Roke C, Codd MB, et al. Hospital mortality of acute myocardial infarction in the thrombolytic era. Heart 1999;81:478-82.

Maier S, Tertilt C, Chambron N, Gerauer K, Huser N, Heidecke C-D, Pfeffer K (2001) Inhibition of natural killer cells results in acceptance of cardiac allografts in CD28-/- mice. Nature Med 7:557-562.

Makino S, Fukuda K, Miyoshi S, et al. Cardiomyocytes can be generated from marrow stromal cells in vitro. J Clin Invest 1999, 103:697-705.

McCarthy PM, Rose EA, Macmanus Q, et al. Clinical experience with the Novacor ventricular assist system. J Thorac Cardiovasc Surg. 1991; 102:578-587.

McEwan PE, Gray GA, Sherry L, Webb DJ, Kenyon CJ. Differential effects of angiotensin II on cardiac cell proliferation and intramyocardial perivascular fibrosis in vivo. Circulation 1998;98:2765-2773.

Nelissen-Vrancken H, Debets J, Snoeckx L, Daemen M, Smits J. Time-related normalization of maximal coronary flow in isolated perfused hearts of rats with myocardial infarction. Circulation 1996;93:349-355.

Nidorf SM, Siu SC, Galambos G, Weyman AE, Picard MH. Benefit of late coronary reperfusion on ventricular morphology and function after myocardial infarction. J Am Coll Cardiol 1992;20:307-313.

Ogawa M, Kizumoto M, Nishikawa S, Fujimoto T, Kodama H, Nishikawa SI. Blood 1999;93:1168-1177.

Oncogene, 9, 3389-3396.

Orlic D, Kajstura J, Chimenti S, et al. Bone marrow cells regenerate infarcted myocardium. Nature 2001, 410:701-705.

Oz MC, Argenziano M, Catanese KA, et al. Bridge experience with long-term implantable left ventricular assist devices. Are they an alternative to transplantation? Circulation. 1997; 95:1844-1852.

Pfeffer JM, Pfeffer MA, Fletcher PJ, Braunwald E. Progressive ventricular remodeling in rat with myocardial infarction. Am J Physiol 1991; 260:H1406-14.

Pfeffer MA, Braunwald E, Moye LA, et al. Effect of captopril on mortality and morbidity in patients with left ventricular dysfunction after myocardial infarction. Results of the survival and ventricular enlargement trial. The SAVE investigators. N Engl J Med 1992;327:669-677.

Pitt B, Segal R, Martinez FA, et al. Randomised trial of losartan versus captopril in patients over 65 with heart failure (Evaluation of Losartan in the Elderly Study, ELITE). Lancet 1997;349:747-752.

Pittenger MF, Mackay AM, Beck SC, et al. Multilineage potential of adult human mesenchymal stem cells. Science 1999, 284:143-147.

Pouzet B, Ghostine S, Vilquin J-T, Garcin I, Scorsin M, Hagege AA, Duboc D, Schwartz K, Menasche P (2001) Is skeletal myoblast transplantation clinically relevant in the era of angiotensin-converting enzyme inhibitors? Circulation 104:I223-228.

Rafii S, Shapiro F, Rimarachin J, Nachman R, Ferris B, Weksler B, Moore AS, Asch AS. Isolation and characterization of human bone marrow microvascular endothelial cells: hematopoietic progenitor cell adhesion.

Ravichandran LV and Puvanakrishnan R. In vivo labeling studies on the biosynthesis and degradation of collagen in experimental myocardial myocardial infarction. Biochem Intl 1991;24:405-414.

Shi Q, Rafii S, Wu MH-D, et al. Evidence for circulating bone marrow-derived endothelial cells. Blood 1998;92:362-367.

Smythe GM, Grounds MD (2000) Exposure to tissue culture conditions can adversely affect myoblast behaviour in vivo in whole muscle grafts: implications for myoblast transfer therapy. Cell Transplant 9:379-393.

Smythe GM, Hodgetts SI, Grounds MD (2001) Problems and solutions in myoblast transfer therapy. J Cell Mol Med 5:33-47.

Soonpaa MH, Field LJ. Assessment of cardiomyocyte DNA synthesis in normal and injured adult mouse hearts. *Am J Physiol* , H220-6 (1997).

Suzuki K, Smolenski RT, Jayakumar J, Murtuza B, Brand NJ, Yacoub MH (2000) Heat shock treatment enhances graft cell survival in skeletal myoblast transplantation to the heart. Circulation 102:III216-221.

Takahashi, T. et al. Ischemia- and cytokine-induced mobilization of bone marrowderived endothelial progenitor cells for neovascularization. Nat. Med. 1999; 5:434-438.

Tavian M, Coulombel L, Luton D, San Clemente H, Dieterlen-Lievre F, Peault B.

Tomita S, Li R-K, Weisel RD, et al. Autologous transplantation of bone marrow cells improves damaged heart function. Circulation 1999; 100:II-247.

Tsai FY, Keller G, Kuo FC, Weiss M, Chen J, Rosenblatt M, Alt FA, Orkin SH. An early hematopoietic defect in mice lacking the transcription factor GATA- 2. Nature 1994;371:221-225.

Tsurimoto, T. PCNA Binding Proteins. Frontiers in Bioscience, 4:849-858, 1999.

White HD, Cross DB, Elliot JM, et al. Long-term prognostic importance of patency of the infarct-related coronary artery after thrombolytic therapy for myocardial infarction. Circulation 1994;89:61-67.

White HD, Norris RM, Brown MA, Brandt PWT, Whitlock RML, Wild CJ. Left ventricular end systolic volume as the major determinant of survival after recovery from myocardial infarction. Circulation 1987; 76:44-51.

Zhang M, Methot D, Poppa V, Fujio Y, Walsh K, Murry CE (2001) Cardiomyocyte grafting for cardiac repair: graft cell death and anti-death strategies. J Mol Cell Cardiol 33:907-921.

2

Biotechnology and Nano World: The Relationship

Some pragmatic readers may be wondering what is so special about molecular nanotechnology, that existing or anticipated biotechnology could not accomplish just as well? After all, biotechnology is already an established medical capability. It has real applications and real products already on the market. Reflecting upon the future possibility of sophisticated mechanical medical nanorobots equipped with powerful nanocomputers, in 1989, one well-known cryobiologist mused that "what is not clear is just what need we would have for such devices." The simplest answer, suggested, is that each of the three contemporary branches of "nanotechnology" offers something of unique value to the practice of medicine.

Nanoscale materials technology has already found widespread use in medicine, including biocompatible materials and analytical techniques, surgical and dental practice, nerve cell research using intracellular electrodes, biostructures research and biomolecular research using near-field optical microscopy, scanning-probe microscopy and optical tweezers, and vaccine design, and also many 20th century bulk chemical and biochemical manufacturing techniques along with much of classical pharmacology. As for "biotechnology," the original meaning of this word contemplated "the application of biological systems and organisms to technical and industrial processes". In recent times, the field has expanded to include genetic engineering and now takes as its ultimate goal no less than the engineering of all biological systems, even completely artificial organic living systems, using biological instrumentalities. The third branch, molecular nanotechnology, takes as its purview the engineering of all complex mechanical systems constructed from the molecular level — potentially offering new tools for medical practice, the principal subject of this book. Observes G.M. Fahy, "the difference between nanotechnologists and biotechnologists is that the former do not restrict themselves to the biological limitations of the latter, and they are much more ambitious about the kinds of accomplishments that they want to achieve." Doctors can utilize solutions to medical problems from all three approaches.

As noted earlier, 80-90% of medical complaints resolve themselves via natural homeostatic processes, or without the necessary involvement of active biotechnological or molecular-nanotechnological agents. But by employing biotechnology, the range and efficacy of treatment options greatly increases. With molecular nanotechnology, the range, efficacy, comfort and speed of possible medical treatments again expands enormously. Molecular nanotechnology is essential when the damage to the human body is extremely subtle, highly selective, or time-critical (as in head traumas, burns, or fast-spreading diseases), or when the damage is very massive, overwhelming the body's natural defenses and repair mechanisms. An earlier discussion from G.M. Fahy, makes these options more explicit. At every difficulty level, most classes of medical problems may be resolved with varying efficacy within the homeostatic/nanomaterials, biotechnological, or molecular-nanotechnological approaches.

As the chosen technology becomes more precise, active, and controllable, the range of options broadens and the quality of the options improves. Thus the question is not whether molecular nanotechnology is required to accomplish a given medical objective. In many cases, it is not — though of course there are some things that only biotechnology and molecular nanotechnology can do, and some other things that only molecular nanotechnology can do.

Rather, the important question is which approach offers a superior solution to a given medical problem, using any reasonable metric of treatment efficacy. For virtually every class of medical challenge, a mature molecular nanotechnology offers a wider and more effective range of treatment options than any other approach. It is quite possible to imagine an advanced biotechnology that uses an engineered white cell, fibroblast, or macrophage chassis, energized by native oxygen and glucose and modified mitochondrial powerplants, driven by pseudopodia, cilia or flagella, communicating and navigating via biochemical signals, and even incorporating onboard digital biocomputers to make microscopic biorobots. Principal arguments favoring the biotechnology approach for medical purposes are: (1) that we are already somewhat familiar with such systems, after half a century of intensive molecular biology research; (2) that we have already "built" precursor systems, such as a whole living amoeba constructed from five distinct parts, bioengineered viruses and bacteria as DNA insertion devices, and natural replication stimulated in genetically engineered starter microbes; (3) that biocompatibility will not be a major issue since fibroblasts (which express no HLA Class II antigens, hence stimulate no rejection response; could be used as the starter material; (4) that both engineered viruses and bacteria are already in wide commercial and research use; and (5) the greater complexity of self-repair in mechanical systems, should it be needed. Many believe that the development pathway to early biorobots may be considerably shorter than for the mechanical nanorobots of the molecular nanotechnology approach, for which in 1998 not a single working prototype yet existed, even in research laboratories. It is also possible to imagine a molecular nanotechnology that uses mechanical nanorobotic systems. Such systems will have many constitutional differences from biological-based systems. For instance, mechanical systems will transport parts, materials, energy and instructions via fixed channels, whereas most (but not all) biological systems operate by diffusion. Mechanical systems will have structures constrained by specific geometries, whereas biological systems have structures defined by patterns of containment and interconnection — the shape of a membrane compartment in a cell matters less than its continuity and the contents of the volume it defines. Mechanical systems will be deterministically manufactured by operations analogous to manual construction, whereas engineered ribosomes will self-assemble via diffusion and stochastic matching of complementary parts; in other words, biology uses recipes, while mechanical systems use blueprints. Cells grow, with their parts adapting to one another; mechanical nanorobots may be constructed from parts of fixed structure.

Biology uses self-repair; mechanical systems generally do not — largely because self-repair (by component exchange) will not be needed in molecular mechanical systems, whose designs may be made more simple by relying upon high component redundancy. These and other differences imply a number of important advantages that mechanical-based medical systems will enjoy over biological-based medical systems, which, taken together, strongly suggest that predominantly mechanical nanosystems may be the approach of choice for a mature medical nanotechnology. The advantages of molecular nanotechnology (e.g. the mechanical tradition) are many:

Speed of Medical Treatment

Doctors may be surprised by the incredible quickness of nanorobotic action when compared to the speeds available from fibroblasts or leukocytes. Normal homeostatic processes such as dermal wound repair via natural fibroblasts may require weeks to run to completion. Typical fibroblast movements occur at 0.1-1 microns/sec, but mechanical nanomanipulators can operate at 1-10 cm/sec speeds or faster, a speed advantage of 4-5 orders of magnitude. Even the strongest biological fibers (e.g. intermediate filaments) have a failure strength 3 orders of magnitude below the strongest mechanical fibers. Biological cilia beat at ~30 Hz while mechanical nanocilia may cycle up to ~20 MHz, though practical power restrictions and other considerations may limit them to the ~10 KHz range for most of the time. Flexible mechanical surfaces can complete a morphing motion in ~0.1 millisec, compared to the ~100 millisec snapback time for pinched red cell membrane, again a thousandfold advantage in speed. Thus we expect that mechanical therapeutic systems can reach their targets up to ~10,000 times faster, all else equal, and treatments which require ~10^5 sec for a biological system may need only ~10^2 sec for a mechanical system; tachyiatria improves both patient and physician comfort. In either biological or mechanical systems, large numbers of devices of comparable physical size (e.g. ~microns) may be employed to do the work, so numbers alone cannot offset the mechanical speed advantage.

Power Density and Transduction

Biological cells typically employ power densities of 10^3-10^4 W/m^3, with maximum densities of ~10^6 W/m^3 in honeybee flight muscle cells and bacterial flagellar motors. By contrast, nanomechanical power systems can produce power densities of 10^9-10^{12} W/m^3, an advantage of 10^3-10^8 for mechanical over biological systems. By 1998, conducting polymer-based actuators generated 20-100 times the force for a given cross-sectional area as mammalian skeletal muscle. Additionally, amoebic locomotion in motile cells requires diffusion-limited cytoskeletal disassembly and reassembly to achieve movement; mechanical motility systems may employ simple cable-pulling, winches, or ratchets, which are faster and more direct. Some biological energy transducers are reversible, but muscle contraction is irreversible — not only cannot muscle actively re-expand, but stretching it doesn't make it produce much useful chemical energy. In contrast, electric motors can be run backwards to generate electricity, forcing pistons makes them pump, and loudspeakers can be used as microphones.

Superior Building Materials

Typical biological materials have tensile failure strengths in the 10^6-10^7 N/m^2 range, with the strongest biological materials such as wet compact bone having a failure strength of ~10^8 N/m^2, all of which compare poorly to ~10^9 N/m^2 for good steel, ~10^{10} N/m^2 for sapphire, and ~10^{11} N/m^2 for diamond and carbon fullerenes, again showing a 10^3-10^5 advantage for mechanical systems that use nonbiological materials. Nonbiological materials can be much stiffer, permitting the application of higher forces with greater precision of movement, and they also tend to remain stable over a larger range of temperature, pressure, salinity and pH. Proteins are heat sensitive in part because much of the functionality of their structure is due to noncovalent bonds involved in folding, which are broken more easily at higher temperatures; in diamond, sapphire, and many other rigid materials, structural shape is covalently fixed, hence is far more temperature-stable. Most proteins tend to become dysfunctional at cryogenic temperatures, unlike diamond-based mechanical structures. Biomaterials are not ruled out for all nanomechanical systems, but represent only a small subset of the materials that can be used in nanorobots. Mechanical systems can employ a wider variety of atoms and molecular structures in their design and construction, with novel functional forms that might be difficult to implement in a biological system such as steam engines or nuclear power. As another example, an application requiring the most effective bulk thermal conduction possible should use diamond, the best conductor available, not some biomaterial with inferior thermal performance.

Nondegradation of Treatment Agents

Diagnostic and therapeutic agents constructed of biomaterials generally are biodegradable in vivo, although there is a major branch of pharmacology devoted to designing drugs that are moderately non-biodegradable — anti-sense DNA analogues with unusual backbone linkages and peptide nucleic acids (PNAs) are difficult to break down. However, suitably designed nanorobotic agents constructed of nonbiological materials are not biodegradable. An engineered fibroblast may not stimulate an immune response when transplanted into a foreign host, but its biomolecules are subject to chemical attack in vivo by free radicals, acids, and enzymes. Even "mirror" biomolecules or "Doppelganger proteins" comprised exclusively of unnatural D-amino acids have a lifetime of only ~5 days inside the human body. Nonbiological materials such as diamond and sapphire are highly resistant to chemical breakdown or leukocytic degradation in vivo, and pathogenic biological entities cannot easily evolve useful attack strategies against these materials.

Control of Nanomedical Treatment

Present-day biotechnological entities are not

programmable and cannot be switched on and off conditionally during task execution. A digital biocomputer, while possible in theory, represents a considerable conceptual departure from the usual biological paradigm. Even assuming that a digital biocomputer could be installed in a fibroblast, and that appropriate effector mechanisms could be attached, such a system would necessarily have slower clock cycles, less capacious memory per unit volume, and longer data access times, implying less diversity of action, poorer control, and less complex executable programs than would be available in nanoscale electromechanical computer systems. The mechanical approach emphasizes precise control of action, including control of physical placement, timing, strength, structure, and interactions with other (especially biological) entities. The biological approach emphasizes the use of poorly controlled natural structures, needlessly sacrificing huge blocks of the available functionality and design space.

Nanodevice Versatility

Mechanical systems can readily incorporate biological elements if necessary, but artificial biological systems can incorporate nonbiological materials such as carbon nanotubes or diamond/sapphire structural elements only with difficulty, in part because biology has a more limited repertoire of "effector" mechanisms. Artificial biological systems cannot easily incorporate nonbiological materials where desired because natural biological assembly methods make no provisions for these materials either in the coded instructions in DNA or in the attachment chemistries. Rebuilding or reconstructing the human body with nonbiological components, or augmentation of human body function with unnatural abilities or features, will be very difficult or impossible to achieve using purely biotechnological means.

Avoiding Overspecialization

M. Krummenacker notes that one of the most glaring shortcomings of bacteria and other naturally occurring molecular machinery — when viewed as systems subject to further engineering — is the rather limited range of molecular substrates they can utilize. For example, bacterial enzymes are highly specialized devices, with very narrow substrate specificities. Thousands of different enzymes are needed in each organism, and the substance classes capable of digestion are limited to some sugars, various amino acids including proteins that can be degraded by excreted proteases, lipids, and a few other smallish oxygen-functionalized carbon molecules such as glycerol and ethanol. Some bacteria can metabolize CO_2 and a handful of aromatic compounds, but there is a vast range of organic chemicals that most bacteria cannot degrade or manufacture. A much smaller set of substantially more general molecular tools can probably be designed using the mechanical approach; mechanosynthesis can fabricate and assemble, or disassemble, a far wider range of molecular structures than are available to the cellular machinery of life.

Faster and More Precise Diagnosis

The analytic function of medical diagnosis requires rapid communication between the injected devices and the attending physician. If limited to chemical messaging, biotechnology devices will require minutes or hours to complete each diagnostic loop. Nanomachines, with their more diverse set of input-output mechanisms, can outmessage the results of in vivo reconnaissance or testing literally in seconds. Such nanomachines can also run more tests of greater variety in less time. Mechanical nanoinstruments, including molecule-by-molecule disassemblers, will make comprehensive cell mapping and cell interaction analysis possible. Bacterial resistance can be assayed at the molecular level, allowing new treatment agents to more easily be composed, manufactured and immediately deployed.

More Sensitive Response Threshold for High-Speed Action

Unlike natural systems, an entire population of nanobiotic devices can be triggered globally by just a single local detection of the target antigen or pathogen. The natural immune system takes $>10^5$ sec to become fully engaged after exposure to a systemic pathogen or other antigen-presenting intruder. A biotechnologically enhanced immune system that can employ the fastest natural unit replication time ($\sim 10^3$ sec for some bacteria) will require $\sim 10^4$ sec for full deployment post-exposure. By contrast, a nanobiotic immune system can probably be fully engaged (though not finished) in at most two blood circulation times, or $\sim 10^2$ sec.

More Reliable Operation

Engineered macrophages would probably individually operate less reliably than mechanical nanorobots. Many pathogens, such as *Listeria monocytogenes* and *Trypanosoma cruzi*, are known to be able to escape from phagocytic vacuoles into the cytoplasm; while biotech drugs or cell manufactured proteins could be developed to prevent this (e.g. cold therapy drugs are entry-point blockers), nanorobotic trapping mechanisms can be more secure. Proteins assembled by natural ribosomes typically incorporate one error per $\sim 10^4$ amino acids placed; current gene and protein synthesizing machines utilizing biotechnological processes have similar error rates. A molecular nanotechnology approach will improve error rates by at least a millionfold. Mechanical systems can also incorporate sensors to determine if and when a particular task needs to be done, or when a task has been completed. Finally, it is unlikely that natural organisms will be able to infiltrate mechanical nanorobots or to co-opt their functions. By contrast, a biological-based robot could be diverted or defeated by microbes that can piggyback on its metabolism, interfere with its normal workings, or even incorporate the device wholesale into their own structures, causing the engineered biomachine to perform some new or different function than was originally intended. There are many examples of such co-option among natural biological systems, including the protozoan mixotrichs found in the termite gut that have assimilated bacteria into their bodies for use as motive engines, and the nudibranch mollusks (marine snails without shells) that steal nematocysts (stinging cells) away from coelenterates such as jellyfish (i.e. a Portuguese man-of-war) and incorporate the stingers as defensive armaments in their own skins, a process which S. Vogel has called "stealing loaded guns from the army."

Verification of Progress and Treatment

Using a variety of communication modalities, nanorobots can report back to the attending physician, with digital precision, a summary of diagnostically- or therapeutically-relevant data describing exactly what was found, and what was done, and what problems were encountered, in every cell visited. A biological-based approach relying upon chemical messaging is necessarily slow with limited signaling capacity. Also unlike mechanical nanorobots, biotechnological systems generally cannot monitor their own functions while working, and, except for a few highly specialized DNA proofreading systems, cannot directly inspect their work while it is in progress or after it is finished.

Minimum Side Effects

Almost all drugs have side effects, such as conventional cancer chemotherapy which causes hair loss and vomiting, although computer-designed drugs have high specificity and relatively few side effects. Carefully tailored cancer vaccines under development in the late 1990s were expected unavoidably to affect some healthy cells. Even well-targeted drugs are distributed to unintended tissues and organs in low concentrations, although some bacteria can target certain organs fairly reliably without being able to distinguish individual cells. By contrast, mechanical nanorobots may be targeted with virtually 100% accuracy to specific organs, tissues, or even individual cellular addresses within the human body. Such nanorobots should have few if any side effects, and will remain safe even in large dosages because their actions can be digitally self-regulated using rigorous control protocols that affirmatively prohibit device activation unless all necessary preconditions have been, and continuously remain, satisfied. G.M. Fahy has noted that these possibilities could transform "drugs" into "programmable machines with a range of sensory, decision-making, and effector capabilities [that] might avoid side effects and allergic reactions...attaining almost complete specificity of action....Designed smart pharmaceuticals might activate themselves only when, where, and if needed." Additionally, nanorobots may be programmed to excuse themselves from the site of action, or even from the body, after a treatment is completed; by contrast, spent biorobotic elements containing ingested foreign materials may have more limited post-treatment mobility, thus lingering at the worksite causing inflammation when naturally degraded or removed R. Bradbury notes that artificial eukaryotic biorobots may possess an apoptotic pathway which could be activated to permit clean and natural self-destruction that avoids inflammation in surrounding tissues. Artificial prokaryotic biorobots could also be designed for human biocompatibility; by replacing bacterial genes with generic human genes, such a device may be undetectable to the immune system.

Reduced Replicator Danger

Drexler points out that living systems are evolved systems, while nanomechanical replicators would be designed: "The former are shaped to serve the goal of their own survival and replication in a natural environment, whereas the latter will be shaped (whether well or poorly) to serve human goals, perhaps in an artificial environment." Genetic engineering involves not design of replicators from scratch, but tinkering with the molecular machinery of existing bioreplicators. Since bioreplicators were not designed, they are not necessarily structured in ways that lend themselves to complete understanding, and processes based on diffusion and matching allow complex nonlocal interactions that can be hard to trace. Bioreplicators can be crippled, but having evolved in nature, they resemble systems that can survive in nature. Typically, they are able to exchange genetic information with wild organisms, raising the possibility of the introduction of new, unconstrained replicators in the natural environment. Having evolved to evolve, they have a capacity for further evolution — to serve their own survival, not human goals. For example, even when stripped of key pieces of DNA to interfere with its replication powers, a live attenuated AIDS vaccine can slowly recover its virulence and can attack immune cells. R. Bradbury suggests that artificial biorobots could incorporate multiple fail-safe mechanisms including required external essential nutrients or suicide suppressors, self-destruct triggers, countdown timers like telomeres, and engineering to reject foreign DNA (the basis for restriction enzymes), but it remains logically easier for a system with inchoate capacity to evolve to resume doing so, than for a system which has never had this capability to spontaneously develop it. In contrast, nanomechanical replicators will be designed from scratch and thus will differ fundamentally from biological systems. The parts and structures of designed mechanical systems will be known, and the relationships among their parts will also be designed and fixed. More important, nanoreplicators will be fundamentally alien to the biosphere, unrelated to anything that has evolved to survive in nature. Certainly the capacity to fail can appear by accident, and emergent capabilities cannot be completely ruled out, but engineering experience shows that the ability to perform complex organized activities (such as replication in a natural environment) does not normally appear spontaneously. Also, and purely as a geometrical consideration, adding a new part inside a densely organized geometric structure (as would be found in a nanomachine) typically requires changes in the relative positions of many other parts, and hence corresponding adjustments in design. Adding a part inside a densely organized topological structure (as found in living systems) typically leaves topologies unchanged — room can be made by stretching and shifting other parts, with no change in their essential design, hence permitting easier modification to biological structures by exogenous agencies. Note that mechanical medical nanodevices need not be capable of replication. There is no requirement for replication in vivo; such replication would be needlessly dangerous, and adding this capability would reduce effectiveness in carrying out the primary medical task. Analogously, viral vectors employed in genetic therapies are modified to be "incapable" of replication.

Assured Patentability

Microscopic biorobots, unavoidably derived from natural biological material, may someday be deemed unpatentable under a general prohibition on "genetic colonialism" or other emerging legal doctrines. In contrast, mechanical nanorobots, being fully-artificial and designed machines, should always be patentable provided they satisfy the customary legal criteria. On the 25th anniversary of Diamond vs. Chakrabarty, the US Supreme Court's landmark decision (June 16, 1980) that opened the floodgates to the patenting of living organisms, ETC Group releases a new report, "Nanotech's 'Second Nature' Patents." Since Chakrabarty, the biotech industry has worked hand-in-hand with governments to allow for the patenting of all biological products – the first monopoly grab over life. Chakrabarty set the stage for today's nanotechnology patents, where the reach of exclusive monopoly is not just on life – but the building blocks of life – nanotech's 'second nature' patents," explains Hope Shand, Research Director of ETC Group. ETC Group's new report examines current trends in intellectual property and nanotechnology and the implications for the developing world. Nanotechnology refers to the manipulation of matter at the scale of atoms and molecules, where size is measured in billionths of meters. The world's largest transnationals, leading academic labs and nanotech start-ups are all racing to win monopoly control of tiny tech's colossal market. "Control and ownership of nanotech is a vital issue for all governments and civil society because nanomaterials and processes can be applied to virtually any manufactured good across all industry sectors," said Kathy Jo Wetter of ETC Group. "Patents are being granted that cut across multiple industry sectors – a single nano-scale innovation may span pharma, food, electronics and materials alike," continues Wetter. The US National Science Foundation predicts that nanotechnology will capture a $1 trillion dollar market within six or seven years. ETC Group finds that breathtakingly broad nanotech patents have been granted that cut across multiple industry sectors and include sweeping claims on entire areas of the Periodic Table. Although industry analysts assert that nanotechnology is in its infancy, "patent thickets" on fundamental nano-scale materials, tools and processes are already creating thorny barriers for would-be innovators. Claims are often broad, overlapping and conflicting—a scenario ripe for massive patent litigation battles in the future. ETC Group's report provides case studies of patent activity involving four of nanotech's hottest and potentially most lucrative nanomaterials and one essential tool: carbon nanotubes; inorganic nanostructures; quantum dots; dendrimers; scanning probe microscopes.

Downsizing Development

When the G8 Summit mets in Scotland, the leaders of the world's most powerful countries unveil of a "Pro-Poor Science" strategy to turn new technologies like nanotech into a silver bullet for social injustice. "Despite rosy predictions that nanotech will provide a technical fix for hunger, disease and the environment, the extraordinary pace of nanotech patenting suggests that developing nations will participate primarily via royalty payments," said Pat Mooney, Executive Director of ETC Group. "In a world dominated by proprietary science, researchers in the global South are likely to find that participation in the nanotech revolution is highly restricted by patent tollbooths, obliging them to pay royalties and licensing fees to gain access," said Mooney. "Ultimately, nanotech will profoundly affect the South's economy, regardless of its handling of intellectual property," explains Silvia Ribeiro from ETC Group's Mexico City office. "Nano-scale technologies will revolutionize the way that new materials are designed and manufactured – changes that could turn commodity markets upside-down and make geography, raw materials, even labour, irrelevant. Nanotech underpins a new strategic platform for global control of materials, food, agriculture and health, and patent monopoly is a powerful tool for realizing that strategy," said Ribeiro. Many South nations are still grappling with unresolved controversies over biotechnology, but by the end of this year, ready or not, even the world's "least developed" nations who are members of the World Trade

Organization will be obligated by its Trade-Related Aspects of Intellectual Property (WTO-TRIPs) to evaluate and enforce nanotech patents.

Lessons Learned from Diamond v. Chakrabarty

Despite all the hype about Mr. Chakrabarty's oileating microbe and how it would gobble up oil spills, the patented microorganism never worked. Instead of curing environmental ills, the biotech industry has introduced its own contamination problems – unwanted gene flow from genetically modified crops, a particularly serious problem for centres of genetic diversity in the developing world. Unlike 25 years ago, today's nanotech-related patents have not required major rule changes. As a result, many governments are unaware of the nanotech patent rush. ETC Group recommends that the World Intellectual Property Organization (WIPO) initiate a global suspension of patent approvals related to nanotechnology until South governments and countries-in-transition can undertake a full evaluation of their impacts, and until social movements can cooperate with WIPO, the Food and Agriculture Organization (FAO) and the United Nations Conference on Trade and Development (UNCTAD) to examine the impact of nanotech-related intellectual property on monopoly practices, technology transfer and trade. The full text of the 32-page report is available for downloading, free-of-charge, on the ETC Group. The Action Group on Erosion, Technology and Concentration, formerly RAFI, is an international civil society organization headquartered in Canada. The ETC Group is dedicated to the advancement of cultural and ecological diversity and human rights. www.etcgroup.org. The ETC Group is also a member of the Community Biodiversity Development and Conservation Programme (CBDC). The CBDC is a collaborative experimental initiative involving civil society organizations and public research institutions in 14 countries.

The CBDC is dedicated to the exploration of community-directed programmes to strengthen the conservation and enhancement of agricultural biodiversity. Implications for the Global South is Nanotech "Pro-Poor Science" or is the G8 Downsizing Development? On the 25th anniversary of the US Supreme Court decision that opened the floodgates to the patenting of all life forms, ETC Group reports on current trends in intellectual property relating to nanoscale technologies. With nanotechnology, the reach of exclusive monopoly patents is not just on life, but all of nature. Accordingly, ETC Group refers to nanotech's "second nature" patents. Nanotech patent claims offer the South (and society) an advance look at who's likely to own nanotechnology and dominate 21st century commodity markets. Breathtakingly broad nanotech patents are being granted that span multiple industry sectors and include sweeping claims on entire classes of the Periodic Table. Yet, when the G8 nations meet this July they will be unveiling a "Pro-Poor Science" strategy to turn nanotechnology into a silver bullet antidote to social injustice. Is nanotech the solution – or just another big downer for development?

The race is on to win monopoly control of tiny tech's colossal market. The US National Science Foundation predicts that the immensely broad power and scope of nano-scale technologies will revolutionize manufacturing across all industry sectors – capturing a $1 trillion market within six or seven years. Although industry analysts assert that nanotech is in its infancy, "patent thickets" on fundamental nano-scale materials, building blocks and tools are already creating thorny barriers for would-be innovators. Industry analysts warn that, "IP roadblocks could severely retard the development of nanotechnology." Some insist that nano-scale technologies will address the most pressing needs of the South's marginalized peoples. But in a world dominated by proprietary science, it is the patent owners and those who can pay license fees who will determine access and price. Even as South governments are grappling with confusion and controversies over biotechnology, the World Trade Organization's Trade- Related Aspects of Intellectual Property (TRIPs) obligates even "least developed" countries to enforce nanotech patents by 2006. The world's largest transnationals, leading academic labs and nanotech start-ups are all racing in the patent gold rush. Increasingly, universities are licensing on an exclusive basis. Nanotech's "second nature patents" are positioning multinational matter moguls to own and control novel materials, devices and their manufacturing processes. While WIPO ponders a "development agenda" in Geneva, patent offices in Washington, Munich and Tokyo are deciding who will gain exclusive monopoly over a technology that will bring profound changes in demand for raw materials and manufacturing around the globe. Despite rosy predictions that nanotech will provide a technical fix for hunger, disease and environmental security in the South, the extraordinary pace of nanotech patenting suggests that developing nations will participate via royalty payments. Accordingly, national governments must protect their sovereignty and intergovernmental agencies must move quickly to prevent multi-sector monopolies, technology barriers and the formation of a powerful new oligopoly that seeks control over the economy's 'second nature.' Control and ownership of nanotechnology is a vital issue for all governments and civil society because nanomaterials and processes can be applied to virtually any manufactured good across *all* industry sectors. Nanotechnology refers to the manipulation of matter at the scale of atoms and molecules, where size is measured in billionths of meters (one nanometer = one-billionth of a meter). Nanotech isn't a single technology – but a range of technologies converging at the nano-scale – including biotechnology, genomics, neurosciences, robotics and information technologies. For a brief introduction to nanotechnology, please see, *A Tiny Primer on Nano-Scale Technologies* [http://www.etcgroup.org] Worldwide, industry and governments invested more than $10 billion in nanotech R&D last year, with two-thirds coming from corporate and private funds.3 There are an estimated 1200 nanotech start-up companies, half of which are US-based.4 Virtually all Fortune 500 companies invest in nanotech R&D. A recent survey conducted by MIT's *Technology Review* predicts that products involving nanotech will account for over $100 billion by 2008; the US government predicts that the nanotech market will explode to $1 trillion by 2012. Intellectual property (IP) will play a major role in deciding who will capture nanotech's trillion dollar market, who will have access to nano-scale technologies and at what price. At stake is control over innovations that span multiple industry sectors – from electronics, energy, mining and defense to new materials, pharmaceuticals and agriculture.

As the *Wall St. Journal* put it, "companies that hold pioneering patents could potentially put up tolls on entire industries." Even industry insiders admit that current intellectual property trends related to nanotech are chaotic. Many broad patents on nanotech-related materials, tools and processes have been granted too early and too often. In 2002, the US-based industry trade group, Nanotechnology Business Alliance, was already warning in testimony before the US Congress, "...several early nanotech patents are given such broad coverage, the industry is potentially in real danger of experiencing unnecessary legal slowdowns." More recently, nanotech industry analysts observe that the "euphoria for patenting" in the US combined with the US Patent & Trademark Office's inability to handle a flood of patent applications has resulted in "the rejection of valid claims, the issuance of broad and over-lapping claims, and a fragmented and somewhat chaotic IP landscape." The writers warn, "These IP roadblocks could severely retard development of nanotechnology." This report aims to provide information to civil society and South policymakers about current trends in nanotechnology and intellectual property. What is the current IP landscape? What is being patented and by whom? What are the implications for innovation and development in the South? How will intellectual property affect the South's participation in the nanotech revolution? "Like biotechnology in the early 1980s, the IP landgrab mentality pays off for early pioneers. Don't bet Patents on nano-scale materials, tools and processes are a fair measure of the tsunami-like strength of this latest industrial revolution.

Estimates vary on the number of nanotech patents issued since the early 1990s, but all agree that both companies and public sector entities are "rushing to the patent office in record numbers to patent nanotechnology inventions." The lack of uniform definitions for nanotechnology means that identifying the number of nanotechrelated patents granted over the past decade is a very imprecise science. When the US PTO announced it had created a new classification for nanotechnology patents in October 2004 it defined nanotechnology patents narrowly:

Classification 977 includes only those patents 1) whose subject matter is in the scale of approximately 1-100 nanometers in at least one dimension; and 2) that involve materials, structures, devices or systems that have novel properties and functions because of their nano-scale size. Nanotech patent searches often use broad search terms (for example, the prefix "nano"), which can result in exaggerated counts. There is wide consensus, however, that major patent offices worldwide are granting nanotech patents at an extraordinary pace. The chart below illustrates the overall trend, which is similar at the World Intellectual Property Organization's Patent Cooperation Treaty and at the US PTO. In response to the demand for nanotech patents, more than a dozen major law firms in the US have recently established nanotech patent law specialties. Researchers from the University of Arizona and the US National Science Foundation examined nanoscale science and engineering patents at the US Patent & Trademark Office from 1976-2003. 12 They found that 8,630 nanotech-related patents were issued by the US PTO in 2003 alone, an increase of 50% over the previous three years. The top 5 countries represented were: US (5,228 patents), Japan (926), Germany (684), Canada (244) and France (183). The top 5 entities winning nanotech-related patents included four multinational electronic firms and one university: IBM (198 patents), Micron Technologies (129), Advanced Micro Devices (128), Intel (90) and University of California (89). A new report by Lux Research, Inc. identifies far fewer nanotech patents granted by the US PTO. In April 2005 Lux announced that it had identified 3,818 nanotech-related patents issued between 1985-March 2005, with an additional 1,777 patent applications.

ETC Group Takes a Closer Look at Nanotech Patents Granted by the US PTO

In October 2004, the US PTO announced it had created a new classification for nanotechnology patents – Class 977 – which would serve as a cross-reference to help examiners, among others, search prior art. Before Class 977 existed, examiners relied on keyword searches to find relevant information and related patents.15 As defined by the US PTO, nanotechnology patents in Class 977 must meet the following criteria:

- relate to research and technology development in

the length scale of approximately 1-100 nm in at least one dimension

- provide a fundamental understanding of phenomena and materials at the nanoscale and create and use structures, devices, and systems that have size-dependent novel properties and functions.16 The US PTO is now reviewing patents issued before the classification was created and re-classifying those that meet the 977 criteria. According to Bruce Kisliuk, a patent examiner group director, the Patent Office hopes to be caught up reviewing already-issued patents within a year.17 Until the review is complete, the PTO is referring to Class 977 as a "digest." Eventually, Class 977 will be broken down further into sub-classes. As of May 25, 2005, Class 977 included 726 patents issued by the US PTO, dating from May 24, 2005 back to 1980. In its incomplete state, Class 977 should be seen as a sample rather than a comprehensive picture of nano-patenting in the US. (At present, Class 977 contains only six nano-patents dating from the second half of 2004, for example.) Even though the classification is incomplete, the sample is worth looking at because it offers an advantage over more random patent searches (e.g., searching for keywords such as "nano" or "quantum"): we can be reasonably confident that the patents in Class 977 are bona fide nano-patents because they have been reviewed and each was determined to meet the classification criteria – relating to the nanometer scale (1-100 nm) in at least one dimension and exploiting size-dependent properties. A search through Class 977 reveals:

The USPTO has issued nano-patents in the fields of electronics (including cameras, computers and other devices), instruments and tools (including scanning probe microscopes and sensors), pharmaceuticals (including drug delivery), food, agrochemicals, devices, materials (including fullerenes, nanotubes and quantum dots) and processes for creating and engineering nano-sized particles.

- More than 290 different primary patent examiners were assigned to evaluate these Class 977 patents. The extraordinarily large number of examiners raises suspicions that over-lapping and conflicting patents were granted by different examiners at about the same time; it also illustrates the multi-disciplinary and cross-sectoral nature of nano-scale science.
- Out of the 235 Class 977 patents issued in 2003 (the most complete year):
- 112 (48%) are assigned to US-based companies
- 29 (over 12%) are assigned to US universities
- 70 (30%) are assigned to companies based outside the US (of these, 26 are assigned to companies based in Japan, 10 in Germany, 9 in Australia – 8 of which are assigned to one company, Silverbrook Research. Companies from Canada, Venezuela, France, Korea, Belgium, Denmark, Italy, Ireland, Netherlands, Taiwan and Antilles make up the other non-US assignees.)
- 15 (6.5%) are assigned to non-US government entities or national research agencies
- The US government has rights to 31 patents (either because it provided funding for research or through outright ownership), which represents 22% of the patents assigned to US entities.

Class 977 patents issued by the USPTO range from mundane to alarming to other-worldly. One would expect only great things from the Delphi Oracle Corporation and it does not disappoint: Delphi owns patent 6,493,910, "a shoelace with enhanced knot retention and method of manufacture." (It's a nanopatent because the shoelace makes use of nano-scale fumed silica.) On the disturbing end of the spectrum, the US government won patent 5,805,657 back in 1998 for "nuclear fuel elements made from nanophase materials," which makes one wonder in the event of infringement – will there be anyone around to sue? Zyvex Corporation owns patent 6,510,359, a "method and system for self-replicating manufacturing stations." In case anyone devises one of these in the next 20 years, it looks like Zyvex will own it. They claim a "non-biological self replicating manufacturing system" comprising:

- a translating machine capable of translating at least a first surface in relation to at least a second surface, wherein said at least a first surface comprises at least one assembly station thereon and wherein said at least a second surface comprises parts arranged thereon for use in constructing at least one other assembly station; and
- a control system for controlling the operation of said translating machine and for controlling said at least one assembly station to cause said at least one assembly station to construct at least one other assembly station from said parts on said second surface.

Patent Paradox

For most nanotech start-up companies (especially those that operate without products or profits), intellectual property is the most essential asset. In the words of one industry executive, the strength of a nanotech company's patent portfolio is what separates the winners from the losers.19 Consider, for example: Nanogen's stock jumped over 50 percent on the day it announced it had won a patent on a technology to detect genetic variants; NVE Corporation's stock surged 41 percent when it announced its new patent on magnetic random access memory. But the transaction costs of filing for and defending patents are enormous.

Competing in the high-stakes patent game requires not only winning and defending patents, but also using them as bargaining chips to cross-license other proprietary technologies. Intellectual property is one of the biggest expenses for fledgling start-ups, and nanotech industry insiders predict that long and costly patent litigation battles could inflict mortal wounds. Consider, for example:

- In the US, depending on the complexity, claims, and length of the application, filing fees can go as high as $1000.21
- For nanotech start-up companies, patent fees are often a major cost of doing business – sometimes second only to payroll.22 Legal fees for winning a patent in the US can cost $25,000- $30,000, and up to $250,000 for international patent claims.23
- Outside the US, countries charge around $5,000 per year on each pending patent. The translation fees alone to win a patent in Japan are between $12,000 and $20,000.24
- The number of patent lawsuits filed in the US more than doubled over the decade of the 1990s. Based on a survey of intellectual property lawyers in 2000, the cost of defending a large (more than $25 million at risk) patent infringement suit ranges from $2 million to $4.5 million. For cases with less than $1

million at risk, the cost was \$300,000 to \$750,000, or about half the amount in dispute.

- Most shocking of all, it is estimated that only about two percent of all issued patents end up generating more revenue than the cost of obtaining the patent. Among this small group of "profitable" patents, only some will end up being worth the price of litigation.

Many intellectual property experts in the US are predicting that large-scale nanotech patent litigation is inevitable, and it's likely to be ugly. Because of the large number of over-lapping and conflicting patents being granted, nanotech companies must be prepared to vigorously defend their patents in court. It's conventional wisdom that, in most patent battles, it's the largest enterprises – not the most innovative – that will prevail. According to authors Josh Lerner and Adam Jaffe, "the firm with the best lawyers or the greatest capacity to withstand the risk of litigation wins the innovation wars – rather than the company with the brightest scientists or most original, valuable ideas."

"Nobody's making any money right now, so there's no particular reason to fight the battles yet. But a lot of people are coming out with patents, and a lot of claims are certainly overlapping, if not certainly *conflicting. The fights are going to be brutal."* – Matthew Nordan, vice-president of research, nanotech analyst firm, Lux Research Firms can use patents as "strategic weapons to strangle competition." If a larger firm believes that its dominance in a nanotechnology market is threatened, it "can slowly starve their start-up competitors into extinction by waging a protracted battle on the IP front."30 As with biotech in the 1980s and 1990s, nanotech start-ups holding key patents may also become attractive takeover targets because it's less costly for a multinational firm to acquire the company than to litigate in court. In part two of this report ETC Group provides case studies of nanotech and intellectual property, focusing on some of the key nanomaterials or tools that are the subject of multiple (and likely) over-lapping patent claims. The case studies examine: carbon nanotubes, Nanosys Inc.'s patents on inorganic nanomaterials, quantum dots, dendrimers and scanning probe microscopes. What are the major trends in Nanotech-related IP?

Patents on Nanotech's Fundamental Building Blocks and Tools

Stanford University law professor Mark Lemley asserts that nanotechnology "is the first new field in a century in which people started patenting the basic ideas at the outset." In contrast to most other major enabling technologies of the 20th century (such as computer hardware, software, the Internet, and even biotechnology), writes Lemley, the most basic ideas and fundamental building blocks in nanotechnology "are either already patented or may well end up being patented." ETC Group does not share Lemley's view that biotech's fundamental enabling tools were not patented early on – but we do share his view that nanotechnology's most basic ideas and building blocks are being patented early and often. In the nanotech arena, it's not just the opportunity to patent the most basic enabling tools, but the ability to patent the nanomaterials themselves, the products they are used in and the methods of making them. At the US Patent & Trademark Office there are three primary types of patent claims:

1. composition of matter claims (that is, nanomaterials such as nanotubes, nanowires and nanoparticles),
2. device, apparatus or system claims (including, for example, tools used to characterize and control nanomaterials—or devices incorporating nanomaterials),

3. method claims (processes for synthesizing nanomaterials or constructing nano-scale devices).

The CEO of Nanosys, Inc., Larry Bock, explains the once-in-a-lifetime attraction of being able to get in on the ground floor by consolidating fundamental nanotech patents: "One of the key things that got me excited about this whole area three years ago is that from an IP standpoint you could control IP from things like composition of matter to the simplest functional devices to the end application all under one IP portfolio. That doesn't come around very often in one's career." Nanomaterials are chemical elements or compounds less than 100 nm in size. Taking advantage of quantum physics, nanotech companies are engineering novel materials that may have entirely new properties never before identified in nature. The "raw materials" for creating nanomaterials and devices are the chemical elements of the Periodic Table – the building blocks of *everything* – both living and nonliving. Whereas biotechnology patents make claims on biological products and processes – nanotechnology patents may literally stake claim to chemical elements, as well as the compounds and the devices that incorporate them. With nano-scale technologies the issue is not just patents on life – but on all of nature. In short, atomic-level manufacturing provides new opportunities for sweeping monopoly control over both animate and inanimate matter. In essence, patenting at the nano-scale could mean monopolizing the basic elements that make life possible.

Proprietary Periodic Table

Exclusive monopoly patents on chemical elements are not new. Glenn Seaborg, the 1951 Nobel Prize-winning physicist, won US patent #3,156,523 for the chemical element *Americium* (element no. 95 on the periodic table) on November 10, 1964. Seaborg's patent is recognized for having the shortest patent claim on record: "What is claimed is Element 95." Seaborg's second patented element was Curium #96 – US patent # 3,161,462 granted on December 15, 1964. "It is true that one cannot patent an element found in its natural form; however, if you create a *purified form of it that has industrial uses—say, neon—you can certainly secure a patent."* – Lila Feisee, Biotechnology Industry Organization's Director for Government Relations and Intellectual Property

It's Elemental

When Harvard University's Charles Lieber obtained a key patent (US patent 5,897,945) on nano-scale metal oxide nanorods, he didn't claim nanorods composed of a single type of metal—but instead claimed a metal oxide selected from up to 33 chemical elements. Harvard's claims on nanorods include those comprised of titanium, zirconium, hafnium, vanadium, niobium, tantalum, chromium, molybdenum, tungsten, manganese, technetium, rhenium, iron, osmium, cobalt, nickel, copper, zinc, cadmium, scandium, yttrium, lanthanum, a lanthanide series element, boron, gallium, indium, thallium, germanium, tin, lead, magnesium, calcium, strontium, and barium.

In a single patent, Lieber's claims extend to nearly one-third of the chemical elements in the Periodic Table – spanning 11 of the 18 Groups. Patent lawyers have identified Harvard's patent (licensed to Nanosys, Inc.) as one of the top 10 patents that could influence the development of nanotechnology.35 Similarly, a key patent on semiconductor nanocrystals (quantum dots) held by the University of California (licensed to Nanosys, Inc. and Quantum Dot Corp.) claims semiconductor nanoparticles from elements in Groups III-V of the Periodic Table. The claims in US patent number 5,505,928 extend to boron, aluminum, gallium, indium, nitrogen, phosphorus, arsenic, antimony as well as those compound semiconductors that result from combining elements in Groups III-V (such as gallium arsenide).

Cross-industry Patent Claims

Nanotechnology is not only cross-disciplinary, a single nano-scale innovation may have diverse applications that span multiple industry sectors. Mark Lemley of Stanford Law School observes, "a significant number of nanotechnology patentees will own rights not just in the industry in which they participate, but in other industries as well."36 When ETC Group examined the 700+ patents that the US Patent & Trademark Office had identified as nanotechnology patents as of May 25, 2005, it was not surprising to find that the patents had originally been assigned to *all* of the major patent classes – including electricity; human necessities; chemistry/metallurgy; performing operations and transporting; mechanical engineering (lighting, heating, weapons, blasting); physics; fixed construction; textiles and paper.

Masters of the Industrial Universe

The crucial aspect to understand about nano IP is not simply that the patents span a broad range of fields, but that a single invention can be relevant for widely divergent applications. It doesn't come as a big surprise to find an invention that is applicable to related fields (e.g., pharmaceuticals and foods), but consider the following examples from US PTO's Class 977:

- US5,874,029 – University of Kansas, 23 February 1999: Methods for particle micronization and nanonization by recrystallization from organic solutions sprayed into a compressed antisolvent: The invention can be used in the pharmaceutical, food, chemical, electronics, catalyst, polymer, pesticide, explosives, and coating industries, all of which have a need for smalldiameter particles.

- US6,667,099 – Creavis Gesellschaft für Technologie und Innovation mbH, 23 December 2003: Meso-and nanotubes: The invention relates to mesotubes and nanotubes (hollow fibers) having an inner diameter of 10 nm-50 μm and to a method for the production thereof ...*The hollow fibers* are used in separation technology, catalysis, micro-electronics, medical technology, material technology or in the clothing industry.
- US6,641,773 – The USA as represented by the Secretary of the Army, 11 November, 2004: Electro spinning of submicron diameter polymer filaments: An electro spinning process yields uniform, nanometer diameter polymer filaments...*The filament is particularly useful for weaving body* armor, for chemical/biological protective clothing, as a biomedical tissue growth support, for fabricating micro sieves and for microelectronics fabrication. The reason that the same invention can be used inside the human body, in clothing and in computers, as in the third example above, is that at the molecular level biological and non-biological material can be integrated – whether this is a seamless integration is a matter yet to be determined by toxicological research.

Second Nature Patents

While biotech's raw materials are biological, nano-scale technologies involve the manipulation of both living and non-living materials, sometimes in combination. When this is the case, the discipline is known as *nanobiotechnology*. A nanostructured material used inside the body as a bone replacement is one example of nanobiotechnology, but so is a hybrid organism created from living and non-living materials, such as the nano-scale silicon and muscle-tissue hybrid announced by researchers in early 2005.37 Closely related to and sometimes overlapping nanobiotech is the new field of "synthetic biology" in which living systems are built to order and then programmed to perform specific tasks. These, too, often combine biological and non-biological parts. Patents on the products of nanobiotechnology provide the opportunity to monopolize the basic elements that are the building blocks of the entire natural world, bringing a whole new dimension to the notion of "life patenting." The table below provides examples of the possible range of nanobiotechnology and synthetic biology patents recently issued by the US PTO. It includes, for example: hybrid devices combining a nanomaterial and muscle tissue, which generate electrical power and which the inventor has described as "absolutely alive"38 (Montemagno); membranes made from biological and non-biological materials to be used in electricity production or water purification (Montemagno); a method for controlling the properties of semiconductor nanoparticles by creating them with the help of biological material (Belcher); synthetic DNA base pairs that do not occur in nature (Benner); a method for genetically modifying cells by pricking them with carbon nanotube "needles" and injecting foreign DNA (McKnight); a gene switch that uses "switching agents" to control gene expression by turning them on or off.

Role of Public Sector Universities in Nanotech IP

One of the unique features of nanotechnology, according to Stanford Law School professor, Mark Lemley, is that universities and public research foundations hold "a grossly disproportionate share of nanotech patents" that he believes are critically important to downstream nanotech products. In 2004 a patent attorney specializing in nanotechnology identified 10 key patents that he believed could have the greatest impact on the development of nanotechnology. Seven of the 10 patents are owned by universities. Because they conduct basic research, it's not surprising that universities are the early-stage engines for nanotechnology. But unlike early-stage researchers 25 years ago, the new generation of US public researchers has become "extremely aggressive patenters" largely because of the Bayh-Dole Act of 1980 – US legislation designed to encourage technology transfer by permitting universities to patent their federally funded research projects. Before 1980, universities worldwide were granted about 250 US patents per year. By 2003, the number of university-owned patents increased almost 16-fold, to 3,933.41 Because university labs aren't in the business of commercializing products, they try to re-coup their research costs by patenting their employees' early scientific innovations – in the hope of earning royalties or licensing fees. Exclusive licensing is generally the more lucrative deal – and therefore the most appealing to technology transfer offices. In general, universities are acting more and more like businesses. Not only are universities patenting nanotech early and often, they are more frequently licensing their inventions on an exclusive basis. US policymakers who favor Bayh-Dole would argue that universities are benefiting society by transferring science and technology to the private sector for commercialization. But in many cases, consumers end up paying twice – once by paying taxes to support governmentfinanced research, and again when they purchase a new, proprietary technology developed with taxpayer funds. With the emphasis on winning exclusive monopoly patents, the traditional academic culture of open communication and exchange is also undermined and eroded.

From 2003 to early 2005 the *Nanotechnology Law & Business Journal* identified 55 publicly announced nanotech patent license agreements – 20 of which involved a university or public research entity as the licensor. Of the 20 license agreements involving university or research entities as licensor, all but one was granted on exclusive terms (and its terms were not disclosed). The largest single holder of nanotech patents in the world is a Chinese researcher, Yang Mengjun, who is taking ancient Chinese medicinal herbs, reducing them to nano-scale formulations, and claiming exclusive monopoly over the herbs or the process used to nano-size them. He holds over 900 patents on nanoscale versions of traditional Chinese medicinal plants.43 Similar patents are being granted in the US and Europe. For example, the Pacific Corporation (Korea) has won a European patent on nanoscale ginseng for use in cosmetic products.44 The Pacific Corporation claims that an emulsion of ginseng at the nano-scale (reduced to small particles between 50-500nm) allows it to penetrate the skin, exerting an antiaging effect. Patent claims on nano-scale formulations of traditional herbal plants are providing insidious pathways to monopolize traditional resources and knowledge – one more reason why the Convention on Biological Diversity and FAO should address the implications of nanotechnology. Diamond vs. Chakrabarty Remembered June 16, 2005 marks the 25th anniversary of Diamond vs. Chakrabarty, the landmark US Supreme Court decision that opened the floodgates to the patenting of all life forms. The anniversary offers a timely opportunity to examine current trends in intellectual property relating to nano-scale technologies – the world's newest technological wave.

In 1971, Ananda Chakrabarty, an employee of General Electric, applied for a patent on a genetically modified, oil-eating microbe. His patent application was rejected by the US Patent & Trademark Office (US PTO) on the grounds that animate life forms were not patentable. When Chakrabarty won his case on appeal, the PTO Commissioner, Sidney Diamond, took the case to the US Supreme Court.

On June 16, 1980 by a narrow 5-4 margin, the US Supreme Court ruled that Chakrabarty's oil-eating microbe was not a product of nature; living organisms could be seen as human made inventions and are therefore patentable subject matter. An ironic footnote to the saga is that the "invention" didn't work. The monumental importance of the Chakrabarty decision did not register with the Court – or the public – at the time. (Some environmentalists were eager to embrace life patenting if it meant microbes could devour oil spills.) In 1980 the Supreme Court specifically noted that the Chakrabarty decision was a narrow case that would not affect the "future of scientific research."45 The Court got it wrong. According to lawyer and activist Andrew Kimbrell, "The complete failure by the Court to correctly assess the impacts of the Chakrabarty decision may go down as among the biggest judicial miscalculations in the Court's long history." As a result of Chakrabarty, the slippery slope of IP on living organisms became a patent landslide, and a bonanza for the biotech industry. Over the course of a single decade, the US government re-interpreted intellectual property laws to allow for exclusive monopoly control over *all* biological products and processes. After Chakrabarty, the once unthinkable patenting of genes, plants, animals, microorganisms and human genetic material would become common practice in the US – positioning industry and the US government to set the precedent for IP regimes worldwide via the World Trade Organization and through bilateral and regional trade agreements.

"Life Patenting" Timeline

1980 — Diamond v. Chakrabarty – US Supreme Court case establishes precedent for the patenting of living organisms.

1980 — US Bayh-Dole Act allows businesses, universities and non-profit organizations to retain title to patents that result from federally-funded research, and to exclusively license them.

1984 — University of California wins US patent on a cell line developed from cancerous tissue of John Moore, a leukemia patient whose cancerous spleen cells were patented and commercialized without his knowledge.

1985 — US PTO rules that genetically engineered plants, seeds and plant tissue are patentable subject matter.

1986 — Uruguay Round of General Agreement on Tariffs and Trade (GATT) begins with Trade-Related Intellectual Property (TRIPs) on the table.

1987 — US PTO rules that genetically modified animals are patentable subject matter.

1988 — US PTO issues first patent on a living animal – a transgenic mouse. Harvard licenses the "OncoMouse" to Dupont.

1993 — Patent claim by the US Secretary of Commerce on the cell line of a 26-year old Guaymi indigenous woman from Panama. Following worldwide controversy, the US government abandoned its patent claim on the Guaymi cell line in November 1993.

1994 — EPO issues species-wide patent on all genetically modified soybeans to Agracetus (later acquired by Monsanto).

1995 — WTO comes into force at the conclusion of the Uruguay Round of GATT.

1995 — Civil society organizations and social movements in Europe defeat European Patent Directive that aims to harmonize patenting of genetic material within the EU.

1995 — US Supreme Court rules in Asgrow vs. Winterboer that farmers no longer have the right to harvest and re-sell proprietary seed for reproductive purposes (that is, proprietary seed protected by Breeders' Rights – plant variety protection – may be saved only for the purpose of re-planting the farmers' own acreage).

1995 — US PTO issues patent to US National Institutes of Health for an unmodified human cell line from an indigenous person from Papua New Guinea. Due to the international controversy, the US government is forced to disclaim the patent in December 1996.

1998 — European Parliament gives final approval to a controversial biotechnology "patent directive" that aims to harmonize national legislation on the patenting of genetic material within the EU. The Directive creates, for the first time in European history, an explicit legal right to patent higher organisms such as plants and animals.

2002 — Canadian Supreme Court rules against patenting of higher life forms– rejects patenting of genetically modified mouse.

2004 — Canadian Supreme Court (Monsanto v. Schmeiser) affirmed Monsanto's right to prosecute farmers who are found to have proprietary GM crops growing on their land – whether they wanted them or not.

2004 — US Patent & Trademark Office establishes nanotechnology patent class.

Lessons learned from Chakrabarty:

- Historic decisions allowing exclusive monopoly control

of all biological products and processes involved no public input or wider societal debate; these decisions were made by a handful of individuals in the courts and patent offices – not by the US Congress. In essence, it was the courts and not citizens who gave biotech the green light in the US. Similarly, at the international level, intellectual property rules have been crafted by and for a narrow group of corporate interests.

- Following Chakrabarty, the US government's aggressive life patenting policies set the bar for the rest of the world – especially at the World Trade Organization.
- The slope is slippery indeed. The history of patent monopoly demonstrates that patent holders typically seek wider patentability, more expansive scope of patent claims, longer patent terms and greater harmonization of patent rules worldwide.
- For many developing nations the rationale for accepting stronger IP regimes has been the argument that their economies would prosper from increased technology transfers and foreign direct investment. In the case of biotechnology, however, the vast majority of key enabling technologies are proprietary products and processes, tightly concentrated in the hands of multinational gene giants. Under these conditions, stronger levels of IP obligate developing countries to make a massive transfer of resources to the North, in order to acquire licenses for proprietary technologies. A new study by the World Bank concludes that the effects of stronger intellectual property regimes in creating greater trade flows to developing countries are "theoretically ambiguous." The authors conclude, however, that stronger levels

of intellectual property in developing countries are not a factor in spurring high-technology trade flows.

The rallying cry "no patents on life" has become a line in a technological and legal sandstorm. Although the notion of intellectual monopolies can be traced back to early Greece, patents did not come into their own until Britain's Industrial Revolution when the inventors of textile machinery demanded "protection." Recognizing that patents would make technology accessible only to well-heeled manufacturers, smaller enterprises protested. The response: "Don't worry. We only seek to patent the machines we invented."

In the 1920s and 30s, when rose and chrysanthemum breeders demanded intellectual property for their flowers, they argued that it was unfair to grant patents to machine inventors but to deny equal rights to ornamental inventors. Although some were repelled by the idea that living things could be patented, the flower companies replied, "Don't worry. These patents protect only decorative plants – not food crops."

In the 1960s, when plant breeders called upon governments to grant them intellectual property over food crops, they said it was unfair to recognize the minor contributions of ornamental breeders without recognizing the contributions of the breeders of crop varieties. The companies chided their critics by saying, "Don't be alarmed. We just want breeders' rights to protect plant varieties; we're not patenting plants, animals or human genetic material, and we would never stop farmers from saving seed."

In 1980, the Gene Giants won patents on genetically modified microbes. A few years later they applied for patents on plants and animals. When civil society protested, industry responded, "Why all the fuss? If you allow the patenting of micro-organisms, why not plants and lab rats?"

In the 1990s, corporations and governments began to patent genes, snippets of DNA, and entire human cell lines. When indigenous peoples protested, patent offices responded, "Don't worry. Human cell lines are just microorganisms."

Meanwhile, patents made it illegal for farmers to save and re-use proprietary seed. The seed/biotech industry denounced the 12,000-year old right of farmers to save harvested seed as patent infringement.

With the advent of nano-scale technologies, corporations are patenting essential building blocks of all living and non-living things. Industry is redefining life to create hybrid organisms that will take on machine functions. When we tell them they have gone too far, they will reply, "Don't worry. We're all just machines."

NANOTECHNOLOGY

Manufactured products are made from atoms. The properties of those products depend on how those atoms are arranged. If we rearrange the atoms in coal, we get diamonds. If we rearrange the atoms in sand (and add a pinch of impurities) we get computer chips. If we rearrange the atoms in dirt, water and air we get grass. Since we first made stone tools and flint knives we have been arranging atoms in great thundering statistical herds by casting, milling, grinding, chipping and the like. We've gotten better at it: we can make more things at lower cost and greater precision than ever before. But at the molecular scale we're still making great ungainly heaps and untidy piles of atoms. That's changing. In special cases we can already arrange atoms and molecules exactly as we want. Theoretical analyses make it clear we can do a lot more. Eventually, we should be able to arrange and rearrange atoms and molecules much as we might arrange LEGO blocks. In not too many decades we should have a manufacturing technology able to:

- Build products with almost every atom in the right place.
- Do so inexpensively.
- Make most arrangements of atoms consistent with physical law.

Often called *nanotechnology*, *molecular nanotechnology* or *molecular manufacturing*, it will let us make most products lighter, stronger, smarter, cheaper, cleaner and more precise.

The Advantages of Nanotechnology

One of the basic principles of nanotechnology is positional control. At the macroscopic scale, the idea that we can hold parts in our hands and assemble them by properly positioning them with respect to each other goes back to prehistory: we celebrate ourselves as the tool using species. Our wisdom and our knowledge would have done us scant good without an opposable thumb: we'd still be shivering in the bushes, unable to start a fire. At the molecular scale, the idea of holding and positioning molecules is new and almost shocking. However, as long ago as 1959 Richard Feynman, the Nobel prize winning physicist, said that nothing in the laws of physics prevented us from arranging atoms the way we want: "...it is something, in principle, that can be done; but in practice, it has not been done because we are too big." What would it mean if we could inexpensively make things with every atom in the right place?

- For starters, we could continue the revolution in computer hardware right down to molecular gates and wires — something that today's lithographic methods (used to make computer chips) could never hope to do.

- We could inexpensively make very strong and very light materials: shatterproof diamond in precisely the shapes we want, by the ton, and over fifty times lighter than steel of the same strength.
- We could make a Cadillac that weighed fifty kilograms, or a full-sized sofa you could pick up with one hand.
- We could make surgical instruments of such precision and deftness that they could operate on the cells and even molecules from which we are made — something well beyond today's medical technology.

The list goes on — almost any manufactured product could be improved, often by orders of magnitude. Nanotechnology should let us make almost every manufactured product faster, lighter, stronger, smarter, safer and cleaner. We can already see many of the possibilities as these few examples illustrate. New products that solve new problems in new ways are more difficult to foresee, yet their impact is likely to be even greater. Could Edison have foreseen the computer, or Newton the communications satellite?

1. Improved Transportation

- Today, most airplanes are made from metal despite the fact that diamond has a strength-to-weight ratio over 50 times that of aerospace aluminum. Diamond is expensive, we can't make it in the shapes we want, and it shatters. Nanotechnology will let us inexpensively make shatterproof diamond (with a structure that might resemble diamond fibers) in exactly the shapes we want. This would let us make a Boeing 747 whose unloaded weight was 50 times lighter but just as strong.
- Today, travel in space is very expensive and reserved

for an elite few. Nanotechnology will dramatically reduce the costs and increase the capabilities of space ships and space flight. The strength-to-weight ratio and the cost of components are absolutely critical to the performance and economy of space ships: with nanotechnology, both of these parameters will be improved... Beyond inexpensively providing remarkably light and strong materials for space ships, nanotechnology will also provide extremely powerful computers with which to guide both those ships and a wide range of other activities in space.

2. Atom Computers

- Today, computer chips are made using lithography — literally, "stone writing." If the computer hardware revolution is to continue at its current pace, in a decade or so we'll have to move beyond lithography to some new post lithographic manufacturing technology. Ultimately, each logic element will be made from just a few atoms.
- Designs for computer gates with less than 1,000 atoms have already been proposed — but each atom in such a small device has to be in exactly the right place. To economically build and interconnect trillions upon trillions of such small and precise devices in a complex three dimensional pattern we'll need a manufacturing technology well beyond today's lithography: we'll need nanotechnology.
- With it, we should be able to build mass storage devices that can store more than a hundred billion billion bytes in a volume the size of a sugar cube; RAM that can store a mere billion billion bytes in such a volume; and massively parallel computers of the same size that can deliver a billion billion instructions per second.

3. Military Applications

- Today, “smart” weapons are fairly big — we have the “smart bomb” but not the “smart bullet.” In the future, even weapons as small as a single bullet could pack more computer power than the largest supercomputer in existence today, allowing them to perform real time image analysis of their surroundings and communicate with weapons tracking systems to acquire and navigate to targets with greater precision and control.
- We'll also be able to build weapons both inexpensively and much more rapidly, at the same time taking full advantage of the remarkable materials properties of diamond. Rapid and inexpensive manufacture of great quantities of stronger more precise weapons guided by massively increased computational power will alter the way we fight wars. Changes of this magnitude could destabilize existing power structures in unpredictable ways. Military applications of nanotechnology raise a number of concerns that prudence suggests we begin to investigate before, rather than after, we develop this new technology.

4. Solar Energy

- Nanotechnology will cut costs both of the solar cells and the equipment needed to deploy them, making solar power economical. In this application we need not make new or technically superior solar cells: making inexpensively what we already know how to make expensively would move solar power into the mainstream.

5. Medical Uses

- It is not modern medicine that does the healing, but the cells themselves: we are but onlookers. If we had surgical tools that were molecular both in their size and precision, we could develop a medical technology that for the first time would let us directly heal the injuries at the molecular and cellular level that are the root causes of disease and ill health. With the precision of drugs combined with the intelligent guidance of the surgeon's scalpel, we can expect a quantum leap in our medical capabilities.

The single most frequently asked question about nanotechnology is: How long? How long before it will let us make molecular computers? How long before inexpensive solar cells let us use clean solar power instead of oil, coal, and nuclear fuel? How long before we can explore space at a reasonable cost? The scientifically correct answer is: *I don't know*. From relays to vacuum tubes to transistors to integrated circuits to Very Large Scale Integrated circuits (VLSI) we have seen steady declines in the size and cost of logic elements and steady increases in their performance.

- Extrapolation of these trends suggests we will have to develop molecular manufacturing in the 2010 to 2020 time frame if we are to keep the computer hardware revolution on schedule.
- Of course, extrapolating past trends is a philosophically debatable method of technology forecasting. While no fundamental law of nature prevents us from developing nanotechnology on this schedule (or even faster), there is equally no law that says this schedule will not slip.
- Much worse, though, is that such trends imply that there is some ordained schedule — that nanotechnology will appear regardless of what we do or don't do. Nothing could be further from the truth. How long it takes to develop this technology

depends very much on what we do. If we pursue it systematically, it will happen sooner. If we ignore it, or simply hope that someone will stumble over it, it will take much longer. And by using theoretical, computational and experimental approaches together, we can reach the goal more quickly and reliably than by using any single approach alone.

While some advances are made through serendipitous accidents or a flash of insight, others require more work. It seems unlikely that a scientist would forget to turn off the Bunsen burner in his lab one afternoon and return to find he'd accidentally made a Space Shuttle. Like the first human landing on the moon, the Manhattan project, or the development of the modern computer, the development of molecular manufacturing will require the coordinated efforts of many people for many years. How long will it take? A lot depends on when we start.

3

DNA Microarrays Technology and It's Uses

Microarraways

The proper and harmonious expression of a large number of genes is a critical component of normal growth and development and the maintenance of proper health. Disruptions or changes in gene expression are responsible for many diseases.

With only a few exceptions, every cell of the body contains a full set of chromosomes and identical genes. Only a fraction of these genes are turned on, however, and it is the subset that is "expressed" that confers unique properties to each cell type. "Gene expression" is the term used to describe the transcription of the information contained within the DNA, the repository of genetic information, into messenger RNA (mRNA) molecules that are then translated into the proteins that perform most of the critical functions of cells. Scientists study the kinds and amounts of mRNA produced by a cell to learn which genes are expressed, which in turn provides insights into how the cell responds to its changing needs. Gene expression is a highly complex and tightly regulated process that allows a cell to respond dynamically both to environmental stimuli and to its own changing needs. This mechanism acts as both an "on/off" switch to control which genes are expressed in a cell as well as a "volume control" that increases or decreases the level of expression of particular genes as necessary.

Enabling Technology

Biomedical research evolves and advances not only through the compilation of knowledge but also through the development of new technologies. Using traditional methods to assay gene expression, researchers were able to survey a relatively small number of genes at a time. The emergence of new tools enables researchers to address previously intractable problems and to uncover novel potential targets for therapies. Microarrays allow scientists to analyze expression of many genes in a single experiment quickly and efficiently. They represent a major methodological advance and illustrate how the advent of new technologies provides powerful tools for researchers. Scientists are using microarray technology to try to understand fundamental aspects of growth and development as well as to explore the underlying genetic causes of many human diseases.

DNA Microarrays

Two recent complementary advances, one in knowledge and one in technology, are greatly facilitating the study of gene expression and the discovery of the roles played by specific genes in the development of disease. As a result of the Human Genome Project, there has been an explosion in the amount of information available about the DNA sequence of the human genome. Consequently, researchers have identified a large number of novel genes within these previously unknown sequences. The challenge currently facing scientists is to find a way to organize and catalog this vast amount of information into a usable form. Only after the functions of the new genes are discovered will the full impact of the Human Genome Project be realized. The second advance may facilitate the identification and classification of this DNA sequence information and the assignment of functions to these new genes: the emergence of DNA microarray technology. A microarray works by exploiting the ability of a given mRNA molecule to bind specifically to, or hybridize to, the DNA template from which it originated. By using an array containing many DNA samples, scientists can determine, in a single experiment, the expression levels of hundreds or thousands of genes within a cell by measuring the amount of mRNA bound to each site on the array. With the aid of a computer, the amount of mRNA bound to the spots on the microarray is precisely measured, generating a profile of gene expression in the cell.

> **A microarray is a tool for analyzing gene expression that consists of a small membrane or glass slide containing samples of many genes arranged in a regular pattern.**

Microarrays are a significant advance both because they may contain a very large number of genes and because of their small size. Microarrays are therefore useful when one wants to survey a large number of genes quickly or when the sample to be studied is small. Microarrays may be used to assay gene expression within a single sample or to compare gene expression in two different cell types or tissue samples, such as in healthy and diseased tissue. Because a microarray can be used to examine the expression of hundreds or thousands of genes at once, it promises to revolutionize the way scientists examine gene expression. This technology is still considered to be in its infancy; therefore, many initial studies using microarrays have represented simple surveys of gene expression profiles in a variety of cell types. Nevertheless, these studies represent an important and necessary first step in our understanding and cataloging of the human genome. As more information accumulates, scientists will be able to use microarrays to ask increasingly complex questions and perform more intricate experiments. With new advances, researchers will be able to infer probable functions of new genes based on similarities in expression patterns with those of known genes. Ultimately, these studies promise to expand the size of existing gene families, reveal new patterns of coordinated gene expression across gene families, and uncover entirely new categories of genes. Furthermore, because the product of any one gene usually interacts with those of many others, our understanding of how these genes coordinate will become clearer through such analyses, and precise knowledge of these inter-relationships will emerge. The use of microarrays may also speed the identification of genes involved in the development of various diseases by enabling scientists to examine a much larger number of genes. This technology will also aid the examination of the integration of gene expression and function at the cellular level, revealing how multiple gene products work together to produce physical and chemical responses to both static and changing cellular needs.

DNA Microarrays are small, solid supports onto which the sequences from thousands of different genes are immobilized, or attached, at fixed locations. The supports themselves are usually glass microscope slides, the size of two side-by-side pinky fingers, but can also be silicon chips or nylon membranes. The DNA is printed, spotted, or actually synthesized directly onto the support. The American Heritage Dictionary defines "array" as "to place in an orderly arrangement". It is important that the gene sequences in a microarray are attached to their support in an orderly or fixed way, because a researcher uses the location of each spot in the array to identify a particular gene sequence. The spots themselves can be DNA, cDNA, or oligonucleotides.

An oligonucleotide, or oligo as it is commonly called, is a

short fragment of a single-stranded DNA that is typically 5 to 50 nucleotides long.

Microarray Experiment

One might ask, how does a scientist extract information about a disease condition from a dime-sized glass or silicon chip containing thousands of individual gene sequences? The whole process is based on hybridization probing, a technique that uses fluorescently labeled nucleic acid molecules as "mobile probes" to identify complementary molecules, sequences that are able to base-pair with one another. Each single-stranded DNA fragment is made up of four different nucleotides, adenine (A), thymine (T), guanine (G), and cytosine (C), that are linked end to end. Adenine is the complement of, or will always pair with, thymine, and guanine is the complement of cytosine.

Therefore, the complementary sequence to G-T-C-C-T-A will be C-A-G-G-A-T. When two complementary sequences find each other, such as the immobilized target DNA and the mobile probe DNA, cDNA, or mRNA, they will lock together, or hybridize. Now, consider two cells: cell type 1, a healthy cell, and cell type 2, a diseased cell. Both contain an identical set of four genes, A, B, C, and D. Scientists are interested in determining the expression profile of these four genes in the two cell types.

To do this, scientists isolate mRNA from each cell type and use this mRNA as templates to generate cDNA with a "fluorescent tag" attached. Different tags (red and green) are used so that the samples can be differentiated in subsequent steps. The two labeled samples are then mixed and incubated with a microarray containing the immobilized genes A, B, C, and D. The labeled molecules bind to the sites on the array corresponding to the genes expressed in each cell.

A DNA Microarray Experiment

1. Prepare your DNA chip using your chosen target DNAs.
2. Generate a hybridization solution containing a mixture of fluorescently labeled cDNAs.
3. Incubate your hybridization mixture containing fluorescently labeled cDNAs with your DNA chip.
4. Detect bound cDNA using laser technology and store data in a computer.
5. Analyze data using computational methods.

After this hybridization step is complete, a researcher will place the microarray in a "reader" or "scanner" that consists of some lasers, a special microscope, and a camera. The fluorescent tags are excited by the laser, and the microscope and camera work together to create a digital image of the array. These data are then stored in a computer, and a special program is used either to calculate the red-to-green fluorescence ratio or to subtract out background data for each microarray spot by analyzing the digital image of the array. If calculating ratios, the program then creates a table that contains the ratios of the intensity of red-to-green fluorescence for every spot on the array. For example, using the scenario outlined above, the computer may conclude that both cell types express gene A at the same level, that cell 1 expresses more of gene B, that cell 2 expresses more of gene C, and that neither cell expresses gene D. But remember, this is a simple example used to demonstrate key points in experimental design. Some microarray experiments can contain up to 30,000 target spots. Therefore, the data generated from a single array can mount up quickly.

The Colors of a Microarray

In this schematic:

GREEN represents Control DNA, where either DNA or

cDNA derived from normal tissue is hybridized to the target DNA.

RED represents Sample DNA, where either DNA or cDNA is derived from diseased tissue hybridized to the target DNA.

YELLOW represents a combination of Control and Sample DNA, where both hybridized equally to the target DNA.

BLACK represents areas where neither the Control nor Sample DNA hybridized to the target DNA.

Each spot on an array is associated with a particular gene. Each color in an array represents either healthy (control) or diseased (sample) tissue. Depending on the type of array used, the location and intensity of a color will tell us whether the gene, or mutation, is present in either the control and/or sample DNA. It will also provide an estimate of the expression level of the gene(s) in the sample and control DNA.

Types

There are three basic types of samples that can be used to construct DNA microarrays, two are genomic and the other is "transcriptomic", that is, it measures mRNA levels. What makes them different from each other is the kind of immobilized DNA used to generate the array and, ultimately, the kind of information that is derived from the chip. The target DNA used will also determine the type of control and sample DNA that is used in the hybridization solution.

Changes in Gene Expression Levels

Determining the level, or volume, at which a certain gene is expressed is called microarray expression analysis, and the arrays used in this kind of analysis are called "expression chips". The immobilized DNA is cDNA derived from the mRNA of known genes, and once again, at least in some experiments, the control and sample DNA hybridized to the chip is cDNA derived from the mRNA of normal and diseased tissue, respectively. If a gene is overexpressed in a certain disease state, then more sample cDNA, as compared to control cDNA, will hybridize to the spot representing that expressed gene. In turn, the spot will fluoresce red with greater intensity than it will fluoresce green.

Once researchers have characterized the expression patterns of various genes involved in many diseases, cDNA derived from diseased tissue from any individual can be hybridized to determine whether the expression pattern of the gene from the individual matches the expression pattern of a known disease. If this is the case, treatment appropriate for that disease can be initiated. As researchers use expression chips to detect expression patterns— whether a particular gene(s) is being expressed more or less under certain circumstances—expression chips may also be used to examine changes in gene expression over a given period of time, such as within the cell cycle.

The cell cycle is a molecualr network that determines, in the normal cell, if the cell should pass through its life cycle. There are a variety of genes involved in regulating the stages of the cell cycle. Also built into this network are mechanisms designed to protect the body when this system fails or breaks down because of mutations within one of the "control genes", as is the case with cancerous cell growth. An expression microarray "experiment" could be designed where cell cycle data are generated in multiple arrays and referenced to time "zero". Analysis of the collected data could further elucidate details of the cell cycle and its "clock", providing much needed data on the points at which gene mutation leads to cancerous growth as well as sources of therapeutic intervention. In the same way, expression chips can be used to develop new drugs. For instance, if a certain gene is overexpressed in a particular form of cancer, researchers can use expression chips to see if a new drug will reduce overexpression and force the cancer into remission. Expression chips could also be used in disease diagnosis as well, e.g., in the identification of new genes involved in environmentally triggered diseases, such as those diseases affecting the immune, nervous, and pulmonary/respiratory systems.

Genomic Gains and Losses

DNA repair genes are thought to be the body's frontline defense against mutations and, as such, play a major role in cancer. Mutations within these genes often manifest themselves as lost or broken chromosomes. It has been hypothesized that certain chromosomal gains and losses are related to cancer progression and that the patterns of these changes are relevant to clinical prognosis. Using different laboratory methods, researchers can measure gains and losses in the copy number of chromosomal regions in tumor cells. Then, using mathematical models to analyze these data, they can predict which chromosomal regions are most likely to harbor important genes for tumor initiation and disease progression. The results of such an analysis may be depicted as a hierarchical treelike branching diagram, referred to as a "tree model of tumor progression". Researchers use a technique called microarray Comparative Genomic Hybridization (CGH) to look for genomic gains and losses or for a change in the number of copies of a particular gene involved in a disease state. In microarray CGH, large pieces of genomic DNA serve as the target DNA, and each spot of target DNA in the array has a known chromosomal location. The hybridization mixture will contain fluorescently labeled genomic DNA harvested from both normal (control) and diseased (sample) tissue. Therefore, if the number of copies of a particular target gene has increased, a large amount of sample DNA will hybridize to those spots on the microarray that represent the gene involved in that disease, whereas comparatively small amounts of control DNA will hybridize to those same spots. As a result, those spots containing the disease gene will fluoresce red with greater intensity than they will fluoresce green, indicating that the number of copies of the gene involved in the disease has gone up.

Mutations in DNA

When researchers use microarrays to detect mutations or polymorphisms in a gene sequence, the target, or immobilized DNA, is usually that of a single gene. In this case though, the target sequence placed on any given spot within the array will differ from that of other spots in the same microarray, sometimes by only one or a few specific nucleotides. One type of sequence commonly used in this type of analysis is called a Single Nucleotide Polymorphism, or SNP, a small genetic change or variation that can occur within a person's DNA sequence.

Another difference in mutation microarray analysis, as compared to expression or CGH microarrays, is that this type of experiment only requires genomic DNA derived from a normal sample for use in the hybridization mixture. Once researchers have established that a SNP pattern is associated with a particular disease, they can use SNP microarray technology to test an individual for that disease expression pattern to determine whether he or she is susceptible to (at risk of developing) that disease. When genomic DNA from an individual is hybridized to an array loaded with various SNPs, the sample DNA will hybridize with greater frequency only to specific SNPs associated with that person. Those spots on the microarray will then fluoresce with greater intensity, demonstrating that the individual being tested may have, or is at risk for developing, that disease.

Microarray Applications

Microarray type	*Application*
CGH	Tumor classification, risk assessment, and prognosis prediction
Expression analysis	Drug development, drug response, and therapy development
Mutation/Polymorphism analysis	Drug development, therapy development, and tracking disease progression

NCBI and Microarray Data Management

Why is it necessary to have a uniform system that will manage and provide a disbursement point for microarray data? Consider the amount of data that can potentially be generated using a single microarray chip. Suppose that chip contains 30,000 spots of target DNA. Researchers interpreting the data generated by that chip would need to know the biological identity of each target—what gene is where; the biological properties of the control and sample DNA; the experimental conditions and procedures used in setting up the experiment; and finally, the results. Although experiments such as these will undoubtedly push forward our current understanding of gene expression and regulation, many new challenges are presented in terms of data tracking and analysis. As we have just alluded, microarray technology is one of the most recent and important experimental breakthroughs in molecular biology. Today, proficiency in generating data is fast overcoming the capacity for storing and analyzing it. Much of this information is scattered across the Internet or is not even available to the public. As more laboratories acquire this technology, the problem will only get worse. This avalanche of data requires standardization of storage, sharing, and publishing techniques. To support the public use and dissemination of gene expression data, NCBI has launched the Gene Expression Omnibus, or GEO. GEO represents NCBI's effort to build an expression data repository and online resource for the storage and retrieval of gene expression data from any organism or artificial source. Many types of gene expression data, such as those types discussed in this primer, are accepted and archived as a public dataset.

Developing MAML: Reading Off the Same Platform

Microarray Markup Language, developed by the "MAML" working group of MGED, the Microarray Gene Expression Database, is a first attempt to provide a standard platform for submitting and analyzing the enormous amounts of microarray expression data generated by different laboratories around the world. The goal of this group, which includes NCBI investigators, is to facilitate the adoption of standards for DNA-array experiment annotation and data representation, as well as the introduction of standard experimental controls and data normalization methods.

The underlying goal is to facilitate the establishment of gene expression data repositories, the comparability of gene expression data from different sources, the interoperability of different gene expression databases, and data analysis software. MAML proposes a framework for describing information about a DNA-array experiment and a data format for communicating this information, including details about:

- *Experimental design:* the set of the hybridization experiments as a whole
- *Array design:* each array used and each spot on the array
- *Samples:* samples used, the extract preparation, and labeling
- *Hybridizations:* procedures and parameters

- *Measurements:* images, quantitation, and specifications
- *Controls:* types, values, and specifications

MAML is independent of the particular experimental platform and provides a framework for describing experiments done on all types of DNA arrays, including spotted and synthesized arrays, as well as oligo and cDNA arrays. What's more, MAML provides format to represent microarray data in a flexible way, which allows analysis of data obtained from not only any existing microarray platforms but also many of the possible future variants, including protein arrays. Although the data in GEO are not currently provided in MAML format, it is NCBI's goal to have the data delivered in a number of formats, including MAML, soon to be replaced by a more recent version called MAGEML (MicroArray Gene Expression Markup Language).

The Benefit of GEO and MAML

- By storing vast amounts of data on gene expression profiles derived from multiple experiments using varied criteria and conditions, GEO will aid in the study of functional genomics—the development and application of global experimental approaches to assess gene function
- GEO will facilitate the cross-validation of data obtained using different techniques and technologies and will help set benchmarks and standards for further gene expression studies
- By making the information stored in GEO publicly available, the fields of bioinformatics and functional genomics will be both promoted and advanced
- That such experimental data should be freely accessible to all is consistent with NCBI's legislative mandate and mission: to develop new information technologies to aid in the understanding of fundamental molecular and genetic processes that control health and disease

Microarray Technology: Future Roles in Treating Diseases

Now that you understand the concept behind array technology, picture this: a hand-held instrument that a physician could use to quickly diagnose cancer or other diseases during a routine office visit. What if that same instrument could also facilitate a personalized treatment regimen, exactly right for you? Personalized drugs. Molecular diagnostics. Integration of diagnosis and therapeutics. These are the long-term promises of microarray technology. Maybe not today or even tomorrow, but someday. For the first time, arrays offer hope for obtaining global views of biological processes—simultaneous readouts of all the body's components—by providing a systematic way to survey DNA and RNA variation. NCBI, by continuing its efforts to provide a standard format for microarray data and to provide free, universal access to that data, will help the scientific community in making those promises realities.

Application of Superconductors

An element, inter-metallic alloy, or compound that will conduct electricity without resistance below a certain temperature. Resistance is undesirable because it produces losses in the energy flowing through the material. Once set in motion, electric current will flow forever in a closed loop of superconducting material – making it the closest thing to perpetual motion in nature. Scientists refer to superconductivity as a "microscopic quantum phenomenon". Superconductivity is a phenomenon which occurs in certain materials and is characterized by the absence of electrical resistivity. Until recently, this phenomenon had been restricted to metals and alloys with transition temperatures of less than 23K. In 1986, superconductivity was discovered in a ceramic material. This precipitated an onrush of ceramic based superconductors with transition temperatures as high as 120K. The ceramic-based materials are commonly known as high temperature (HTc) superconductors while the metallic and alloy materials are called low temperature (LTc) superconductors. Currently, only low temperature superconductors are of interest to the magnet designer and manufacturer. Superconductors are divided into two types depending on their characteristic behavior in the presence of a magnetic field. Type I superconductors are comprised of pure metals, whereas Type II superconductors are comprised primarily of alloys or intermetallic compounds. Both, however, have one common feature: below a critical temperature, T_c, their resistance vanishes. The critical temperature at which the resistance vanishes in a superconductor is reduced when a magnetic field is applied. The maximum field that can be applied to a superconductor at a particular temperature and still maintain superconductivity is call the critical field, or H_c. This field varies enormously between Type I and Type II superconductors.

The maximum critical field (H_c) in any Type I superconductor is about 2000 Gauss (0.2 Tesla), but in Type II materials superconductivity can persist to several hundred thousand Gauss (H_{c2}). At fields greater than H_c in a Type I superconductor and greater than H_{c2} in a Type II superconductor, the conductor reverts to the normal state and regains its normal state resistance. A Type I superconductor excludes the applied magnetic field from the center of the sample by establishing circulating currents on its surface that counteract the applied field. Type II superconductors, however, permit the field to penetrate through the sample in quantized amounts of flux. These quanta are comprised of circulating vortices of current and the flux contained in the vortices. The total flux in a vortex is 2×10^{-7} Gauss-cm^2. Great numbers of these vortices, or fluxoids as they are frequently called, can exist in a superconductor. For example, at a field intensity of 80 kilogauss (8 Tesla) there are 4 x 10^{11} fluxoids/cm^2. These fluxoids and their interactions with defects in the superconductor give rise to the high current carrying capabilities of superconducting magnets.

Properties of superconducting materials are altered locally by the presence of defects in the materials. A fluxoid encompassing or adjacent to such a defect in the material has its energy altered and its free motion through the superconductor is inhibited. This phenomenon, known as flux pinning, causes a field gradient in the superconductor and gives rise to a net current in the material. In the absence of defects in a Type II superconductor, no bulk current can be conducted without a transition into the normally conducting resistive state. Since the pinning force is small, fluxoids can be broken loose from their pinning centers, resulting in a net creep of the flux through a conductor as a function of time.

This results in an effective voltage in a Type II superconductor. If the current density is low and the magnetic field is not intense, flux creep is insignificant and the induced voltage and effective resistance of the conductor will be essentially zero. At very high fields and high current densities, fluxoids will migrate rapidly, giving rise to a phenomenon called flux flow. The maximum pinning force and the maximum current density sustainable by a conductor increases with reduced temperature. Similarly, as the magnetic induction, B, increases, the sustainable current density decreases. Consequently, the current density, magnetic field, and critical temperature are all interdependent. By increasing any of these parameters to a sufficiently high value, superconductivity can be destroyed and the conductor will revert to a normal, non-superconducting state. One of the most crucial factors in determining the performance of the final magnet is the design of the conductor. This design affects the ultimate field achieved by the magnet, the rate at which the magnet can be energized and the drift rate in the persistent mode of operation. Several phenomena observed in magnets are caused by the conductor itself.

One of the earliest phenomena observed in superconducting magnets wound with single filament conductors was flux jumping. This phenomena arises from current induced in the conductor by the presence of the transverse field generated by the magnet. If a superconductor is placed transverse to the magnetic field, currents are induced in the conductor which shield the bulk of the conductor from the external magnetic field. These circulating currents extend for a finite length along the conductor, flowing in one direction on one side of the conductor and returning on the other side to complete the circuit. If the heat produced by the penetration of the magnet field into the superconductor cannot escape to the surface rapidly enough, then the temperature increase inside the superconductor triggers a runaway condition known as a flux jump. Consequently, the conductor is driven into the resistive state at fields and currents low in comparison with the critical values. Resistive heat is then dissipated in the small normal zone which increases in temperature causing the normal zone to expand and propagate both along the length of the conductor and transverse to it.

This results in the magnet being discharged as the energy in the magnet is dissipated in the resistive portion of the conductor. To decrease the problem of transition to the normal state, it is common practice to shunt the conductor with a low resistivity normal metal by embedding the superconductor in a copper matrix to form a composite conductor. The copper provides additional heat capacity as well as providing a path for the magnet current while the superconductor is driven normal during a flux jump. If the resistance of the copper is low enough, the temperature of the conductor can remain below the critical temperature at the ambient field, and superconductivity will resume after the currents in the superconductor have decayed. Embedding the superconductor in a low resistivity metal matrix is effective in reducing the chance of a flux jump which can cause a magnet quench. Magnets constructed with this type of material are dissipative and the heat generated during a flux jump must be conducted to the helium bath. Thus, the magnetic field must be changed slowly to allow time for the heat to be conducted to and dissipated in the liquid helium.

Also, the diamagnetic currents in the superconductor contribute to the field generated by the magnet and can reduce its homogeneity. If, instead of one superconducting filament, many fine filaments of superconductor are used, the heat generated in individual filaments can easily be conducted a short distance to the filament surface thus avoiding flux jumps. Consequently, conductors are made in which many fine filaments of superconductor are coextruded and drawn in a matrix of either copper or aluminum stabilizers. Although this has the desired effect of avoiding flux jumping, circulating currents can again be formed if the conductors are parallel in the highly conductive normal matrix. In this case, the circulating current is between two or more filaments in parallel and the current crosses over through the normally conductive matrix. This gives rise to diamagnetism and unequal distribution of currents in the filaments that limits the rate at which the magnet can be charged. Problems arising from constructing the superconductor from filaments have been largely circumvented in modern conductors by twisting the filaments in the conductor. This causes the flux from the external magnetic field to be alternated through successive superconducting filaments, thereby reducing the unequal distribution of currents between the superconducting filaments and reducing the diamagnetism of the conductor.

This reduction in diamagnetism or hysteresis has two desirable effects. First, it reduces the amount of energy dissipated in the magnet and permits it to be charged more rapidly. Secondly, the reduced diamagnetism causes the current in the magnet and the magnetic field generated by the magnet to be more linearly related. Such conductors are known as intrinsically stabilized conductors. Superconductors, materials that have no resistance to the flow of electricity, are one of the last great frontiers of scientific discovery. Not only have the limits of superconductivity not yet been reached, but the theories that explain superconductor behavior seem to be constantly under review. In 1911 superconductivity was first observed in mercury by Dutch physicist Heike Kamerlingh Onnes of Leiden University (shown above). When he cooled it to the temperature of liquid helium, 4 degrees Kelvin (-452F, -269C), its resistance suddenly disappeared.

The Kelvin scale represents an "absolute" scale of temperature. Thus, it was necessary for Onnes to come within 4 degrees of the coldest temperature that is theoretically attainable to witness the phenomenon of superconductivity. Later, in 1913, he won a Nobel Prize in physics for his research in this area. The next great milestone in understanding how matter behaves at extreme cold temperatures occurred in 1933. German researchers Walter Meissner (above) and Robert Ochsenfeld discovered that a superconducting material will repel a magnetic field (below graphic). A magnet moving by a conductor induces currents in the conductor. This is the principle upon which the electric generator operates. But, in a superconductor the induced currents exactly mirror the field that would have otherwise penetrated the superconducting material—causing the magnet to be repulsed. This phenomenon is known as strong diamagnetism and is today often referred to as the "Meissner effect" (an eponym). The Meissner effect is so strong that a magnet can actually be levitated over a superconductive material. In subsequent decades other superconducting metals, alloys and compounds were discovered. In 1941 niobium-nitride was found to superconduct at 16 K. In 1953 vanadium-silicon displayed superconductive properties at 17.5 K. And, in 1962 scientists at Westinghouse developed the first commercial superconducting wire, an alloy of niobium and titanium (NbTi). High-energy, particle-accelerator electromagnets made of copper-clad niobium-titanium were then developed in the 1960s at the Rutherford-Appleton Laboratory in the UK, and were first employed in a superconducting accelerator at the Fermilab Tevatron in the US in 1987.

The first widely-accepted theoretical understanding of superconductivity was advanced in 1957 by American physicists John Bardeen, Leon Cooper, and John Schrieffer (above). Their *Theories of Superconductivity* became know as the BCS theory - derived from the first letter of each man's last name—and won them a Nobel prize in 1972. The mathematically-complex BCS theory explained superconductivity at temperatures close to absolute zero for elements and simple alloys. However, at higher temperatures and with different superconductor systems, the BCS theory has subsequently become inadequate to fully explain how superconductivity is occurring. Another significant theoretical advancement came in 1962 when Brian D. Josephson (above), a graduate student at Cambridge University, predicted that electrical current would flow between 2 superconducting materials - even when they are separated by a non-superconductor or insulator. His prediction was later confirmed and won him a share of the 1973 Nobel Prize in Physics. This tunneling phenomenon is today known as the "Josephson effect" and has been applied to electronic devices such as the SQUID, an instrument capabable of detecting even the weakest magnetic fields.

The 1980's were a decade of unrivaled discovery in the field of superconductivity. In 1964 Bill Little of Stanford University had suggested the possibility of organic (carbon-based) superconductors. The first of these theoretical superconductors was successfully synthesized in 1980 by Danish researcher Klaus Bechgaard of the University of Copenhagen and 3 French team members. $(TMTSF)_2PF_6$ had to be cooled to an incredibly cold 1.2K transition temperature (known as Tc) and subjected to high pressure to superconduct. But, its mere existence proved the possibility of "designer" molecules - molecules fashioned to perform in a predictable way. Then, in 1986, a truly breakthrough discovery was made in the field of superconductivity. Alex Müller and Georg Bednorz (above), researchers at the IBM Research Laboratory in Rüschlikon, Switzerland, created a brittle ceramic compound that superconducted at the highest temperature then known: 30 K. What made this discovery so remarkable was that ceramics are normally insulators. They don't conduct electricity well at all. So, researchers had not considered them as possible high-temperature superconductor candidates. The Lanthanum, Barium, Copper and Oxygen compound that Müller and Bednorz synthesized, behaved in a not-as-yet-understood way. The discovery of this first of the superconducting copper-oxides (cuprates) won the 2 men a Nobel Prize the following year. It was later found that tiny amounts of this material were actually superconducting at 58 K, due to a small amount of lead having been added as a calibration standard - making the discovery even more noteworthy. Müller and Bednorz' discovery triggered a flurry of activity in the field of superconductivity. Researchers around the world began "cooking" up ceramics of every imaginable combination in a quest for higher and higher Tc's. In January of 1987 a research team at the University of Alabama-Huntsville substituted Yttrium for Lanthanum in the Müller and Bednorz molecule and achieved an incredible 92 K Tc.

For the first time a material (today referred to as YBCO) had been found that would superconduct at temperatures warmer than liquid nitrogen - a commonly available coolant. Additional milestones have since been achieved using exotic—and often toxic—elements in the base perovskite ceramic. The current class (or "system") of ceramic superconductors with the highest transition temperatures are the mercuric-cuprates. The first synthesis of one of these compounds was achieved in 1993 by Prof. Dr. Ulker Onbasli at the University of Colorado and by the team of A. Schilling, M. Cantoni, J.D. Guo, and H.R. Ott of Zurich, Switzerland. The world record Tc of 138 K is now held by a thallium-doped, mercuric-cuprate comprised of the elements Mercury, Thallium, Barium, Calcium, Copper and Oxygen. The Tc of this ceramic superconductor was confirmed by Dr. Ron Goldfarb at the National Institute of Standards and Technology-Colorado in February of 1994. Under extreme pressure its Tc can be coaxed up even higher - approximately 25 to 30 degrees more at 300,000 atmospheres. The first company to capitalize on high-temperature superconductors was Illinois Superconductor (today known as ISCO International), formed in 1989. This amalgam of government, private-industry and academic interests introduced a depth sensor for medical equipment that was able to operate at liquid nitrogen temperatures (~ 77K). While no significant advancements in superconductor Tc's have been achieved in recent years, other discoveries of equal importance have been made. In 1997 researchers discovered that at a temperature very near absolute zero an alloy of gold and indium was both a superconductor and a natural magnet. Conventional wisdom held that a material with such properties could not exist!

Since then, over a half-dozen such compounds have been found. Recent years have also seen the discovery of the first high-temperature superconductor that does NOT contain any copper (2000), and the first all-metal perovskite superconductor (2001). Also in 2001 a material that had been sitting on laboratory shelves for decades was found to be an extraordinary new superconductor. Japanese researchers measured the transition temperature of magnesium diboride at 39 Kelvin—far above the highest Tc of any of the elemental or binary alloy superconductors. While 39 K is still well below the Tc's of the "warm" ceramic superconductors, subsequent refinements in the way MgB_2 is fabricated have opened the door to its use in industrial applications. Laboratory testing has found MgB_2 will outperform NbTi and Nb_3Sn wires in high magnetic field applications like MRI. Startling—and serendipitous—discoveries like these are forcing scientists to continually re-examine longstanding theories on superconductivity and to consider heretofore-unimagined combinations of elements. Magnetic-levitation is an application where superconductors perform extremely well. Transport vehicles such as trains can be made to "float" on strong superconducting magnets, virtually eliminating friction between the train and its tracks. Not only would conventional electromagnets waste much of the electrical energy as heat, they would have to be physically much larger than superconducting magnets. A landmark for the commercial use of MAGLEV technology occurred in 1990 when it gained the status of a nationally-funded project in Japan. The Minister of Transport authorized construction of the Yamanashi Maglev Test Line which opened on April 3, 1997. In December 2003, the MLX01 test vehicle (shown above) attained an incredible speed of 361 mph (581 kph). Although the technology has now been proven, the wider use of MAGLEV vehicles has been constrained by political and environmental concerns (strong magnetic fields can create a bio-hazard). The world's first MAGLEV train to be adopted into commercial service, a shuttle in Birmingham, England, shut down in 1997 after operating for 11 years. A Sino-German maglev is currently operating over a 30 km course at Pudong International Airport in Shanghai, China. The U.S. plans to have its first (non-superconducting) Maglev train in operation this year on a Virginia college campus. Click this link for a website that lists other uses for MAGLEV.

An area where superconductors can perform a life-saving function is in the field of biomagnetism. Doctors need a non-invasive means of determining what's going on inside the human body. By impinging a strong superconductor-derived magnetic field into the body, hydrogen atoms that exist in the body's water and fat molecules are forced to accept energy from the magnetic field. They then release this energy at a frequency that can be detected and displayed graphically by a computer. Magnetic Resonance Imaging (MRI) was actually discovered in the mid 1940's. But, the first MRI exam on a human being was not performed until July 3, 1977. And, it took almost five hours to produce one image! Today's faster computers process the data in much less time. The Korean Superconductivity Group within KRISS has carried biomagnetic technology a step further with the development of a double-relaxation oscillation SQUID (Superconducting quantum Interference Device) for use in Magnetoencephalography. SQUID's are capable of sensing a change in a magnetic field over a billion times weaker than the force that moves the needle on a compass (compass: 5e-5T, SQUID: e-14T.). With this technology, the body can be probed to certain depths without the need for the strong magnetic fields associated with MRI's. Probably the one event, more than any other, that has been responsible for putting "superconductors" into the American lexicon was the Superconducting Super-Collider project planned for construction in Ellis county, Texas.

Though Congress cancelled the multi-billion dollar effort in 1993, the concept of such a large, high-energy collider would never have been viable without superconductors. High-energy particle research hinges on being able to accelerate sub-atomic particles to nearly the speed of light. Superconductor magnets make this possible. CERN, a consortium of several European nations, is doing something similar with its Large Hadron Collider (LHC) now under construction along the Franco-Swiss border. Other related web sites worth visiting include the proton-antiproton collider page at Fermilab. This was the first facility to use superconducting magnets. Get information on the electron-proton collider HERA at the German lab pages of DESY (with English text). Lastly, Brookhaven National Laboratory features a page dedicated to its RHIC heavy-ion collider. Electric generators made with superconducting wire are far more efficient than conventional generators wound with copper wire. In fact, their efficiency is above 99 per cent and their size about half that of conventional generators. These facts make them very lucrative ventures for power utilities. General Electric has estimated the potential worldwide market for superconducting generators in the next decade at around $20-30 billion dollars. Late in 2002 GE Power Systems received $12.3 million in funding from the U.S. Department of Energy to move high-temperature superconducting generator technology toward full commercialization. Other commercial power projects in the works that employ superconductor technology include energy storage to enhance power stability.

American Superconductor Corp. received an order from Alliant Energy in late March 2000 to install a Distributed Superconducting Magnetic Energy Storage System (D-SMES) in Wisconsin. Just one of these 6 D-SMES units has a power reserve of over 3 million watts, which can be retrieved whenever there is a need to stabilize line voltage during a disturbance in the power grid. AMSC has also installed more than 22 of its D-VAR systems to provide instantaneous reactive power support. Recently, power utilities have also begun to use superconductor-based transformers and "fault limiters". The Swiss-Swedish company ABB was the first to connect a superconducting transformer to a utility power network in March of 1997. ABB also recently announced the development of a 6.4MVA (mega-volt-ampere) fault current limiter - the most powerful in the world. This new generation of HTS superconducting fault limiters is being called upon due to their ability to respond in just thousandths of a second to limit tens of thousands of amperes of current. Advanced Ceramics Limited is another of several companies that makes BSCCO type fault limiters. Intermagnetics General recently completed tests on its largest (15kv class) power-utility-size fault limiter at a Southern California Edison (SCE) substation near Norwalk, California. And, both the US and Japan have plans to replace underground copper power cables with superconducting BSCCO cable-in-conduit cooled with liquid nitrogen. By doing this, more current can be routed through existing cable tunnels. In one instance 250 pounds of superconducting wire replaced 18,000 pounds of vintage copper wire, making it over 7000 per cent more space-efficient. An idealized application for superconductors is to employ them in the transmission of commercial power to cities. However, due to the high cost and impracticality of cooling miles of superconducting wire to cryogenic temperatures, this has only happened with short "test runs". In May of 2001 some 150,000 residents of Copenhagen, Denmark, began receiving their electricity through HTS (high-temperature superconducting) material. That cable was only 30 meters long, but proved adequate for testing purposes. In the summer of 2001 Pirelli completed installation of three 400-foot HTS cables for Detroit Edison at the Frisbie Substation capable of delivering 100 million watts of power.

This marked the first time commercial power has been delivered to customers of a US power utility through superconducting wire. Intermagnetics General has announced that its IGC-SuperPower subsidiary has joined with BOC and Sumitomo Electric in a $26 million project to install an underground, HTS power cable in Albany, New York, in Niagara Mohawk Power Corporation's power grid. The 350-meter cable, believed to be four times the length of any previously installed HTS cable, will be designed to provide more power and operate at significantly lower loss levels than other HTS installations. The National Science Foundation, along with NASA and DARPA and various universities, are currently researching "petaflop" computers. A petaflop is a thousand-trillion floating point operations per second. Today's fastest computing operations have only reached "teraflop" speeds - trillions of operations per second. Currently the fastest is the IBM Blue Gene/L running at 70.7 teraflops per second (multiple CPU's). The fastest single processor is a Lenslet optical DSP running at 8 teraflops. It has been conjectured that devices on the order of 50 nanometers in size along with unconventional switching mechanisms, such as the Josephson junctions associated with superconductors, will be necessary to achieve such blistering speeds. TRW researchers (now Northrop Grumman) have quantified this further by predicting that 100 billion Josephson junctions on 4000 microprocessors will be necessary to reach 32 petabits per second. These Josephson junctions are incorporated into field-effect transistors which then become part of the logic circuits within the processors. Recently it was demonstrated at the Weizmann Institute in Israel that the tiny magnetic fields that penetrate Type 2 superconductors can be used for storing and retrieving digital information.

It is, however, not a foregone conclusion that computers of the future will be built around superconducting In the electronics industry, ultra-high-performance filters are now being built. Since superconducting wire has near zero resistance, even at high frequencies, many more filter stages can be employed to achive a desired frequency response. This translates into an ability to pass desired frequencies and block undesirable frequencies in high-congestion rf (radio frequency) applications such as cellular telephone systems. ISCO International and Superconductor Technologies are companies currently offering such filters. Superconductors have also found widespread applications in the military. HTSC SQUIDS are being used by the US NAVY to detect mines and submarines. And, significantly smaller motors are being built for NAVY ships using superconducting wire and "tape". In mid-July, 2001, American Superconductor unveiled a 5000-horsepower motor made with superconducting wire. The military is also looking at using superconductive tape as a means of reducing the length of very low frequency antennas employed on submarines. Normally, the lower the frequency, the longer an antenna must be. However, inserting a coil of wire ahead of the antenna will make it function as if it were much longer. Unfortunately, this loading coil also increases system losses by adding the resistance in the coil's wire. Using superconductive materials can significantly reduce losses in this coil. The Electronic Materials and Devices Research Group at University of Birmingham (UK) is credited with creating the first superconducting microwave antenna. Applications engineers suggest that superconducting carbon nanotubes might be an ideal nano-antenna for high-gigahertz and terahertz frequencies, once a method of achieving zero "on tube" contact resistance is perfected.

The most ignominious military use of superconductors may come with the deployment of "E-bombs". These are devices that make use of strong, superconductor-derived magnetic fields to create a fast, high-intensity electromagnetic pulse (EMP) to disable an enemy's electronic equipment. Such a device saw its first use in wartime in March 2003 when US Forces attacked an Iraqi broadcast facility. Among emerging technologies are a stabilizing momentum wheel (gyroscope) for earth-orbiting satellites that employs the "flux-pinning" properties of imperfect superconductors to reduce friction to near zero. Superconducting x-ray detectors and ultra-fast, superconducting light detectors are being developed due to their inherent ability to detect extremely weak amounts of energy. Already Scientists at the European Space Agency (ESA) have developed what's being called the S-Cam, an optical camera of phenomenal sensitivity. And, superconductors may even play a role in Internet communications soon. In late February, 2000, Irvine Sensors Corporation received a $1 million contract to research and develop a superconducting digital router for high-speed data communications up to 160 Ghz. Since Internet traffic is increasing exponentially, superconductor technology is being called upon to meet this super need. According to

June 2002 estimates by the Conectus consortium, the worldwide market for superconductor products is projected to grow to near US $5 billion by the year 2010 and to US $38 billion by 2020. Low-temperature superconductors are expected to continue to play a dominant role in well-established fields such as MRI and scientific research, with high-temperature superconductors enabling the newer industries. The above ISIS graph gives a rough breakdown of the various markets in which superconductors are expected to make a contribution.

All of this is, of course, contingent upon a linear growth rate. Should new superconductors with higher transition temperatures be discovered, growth and development in this exciting field could explode virtually overnight. Another impetus to the wider use of superconductors is political in nature. The reduction of green-house gas (GHG) emissions has becoming a topical issue due to the Kyoto Protocol which requires the European Union (EU) to reduce its emissions by 8 per cent from 1990 levels by 2012. Physicists in Finland have calculated that the EU could reduce carbon dioxide emissions by up to 53 million tons if high-temperature superconductors were used in power plants. The future melding of superconductors into our daily lives will also depend to a great degree on advancements in the field of cryogenic cooling. New, high-efficiency magnetocaloric-effect compounds such as gadolinium-silicon-germanium are expected to enter the marketplace soon. Such materials should make possible compact, refrigeration units to facilitate additional HTS applications. Laboratories throughout the world are rapidly gaining atomically precise control over matter. As this control extends to an ever wider variety of materials, processes and devices, opportunities for applications relevant to NASA's missions will be created. This document surveys a number of future molecular nanotechnology capabilities of aerospace interest. Computer applications, launch vehicle improvements, and active materials appear to be of particular interest. We also list a number of applications for each of NASA's enterprises. If advanced molecular nanotechnology can be developed, almost all of NASA's endeavors will be radically improved. In particular, a sufficiently advanced molecular nanotechnology can arguably bring large scale space colonization within our grasp. This study describes potential aerospace applications of molecular nanotechnology, defined as the thorough three-dimensional structural control of materials, processes and devices at the atomic scale. The inspiration for molecular nanotechnology comes from Richard P. Feynman's 1959 visionary talk at Caltech in which he said, "The problems of chemistry and biology can be greatly helped if our ability to see what we are doing, and to do things on an atomic level, is ultimately developed—a development which I think cannot be avoided." Indeed, scanning probe microscopes (SPMs) have already given us this ability in limited domains. Synthetic chemistry, biotechnology, "laser tweezers" and other developments are also bringing atomic precision to our endeavors.

An expanded version of Drexler's MIT PhD thesis, examines one vision of molecular nanotechnology in considerable technical detail. It proposes the development of programmable molecular assembler/replicators. These are atomically precise machines that can make and break chemical bonds using mechanosynthesis to produce a wide variety of products under software control, including copies of themselves. Interestingly, living cells exhibit many properties of assembler/replicators. Cells make a wide variety of products, including copies of themselves, and can be programmed with DNA. Replication is one approach to building large systems, such as human rated launch vehicles, from molecular machines manipulating matter one or a few atoms at a time. Note that biological replication is responsible for systems as large as redwood trees and whales. Another approach to nanotechnology is supramolecular self-assembly, where molecular systems are designed to attract each other in a particular orientation to form larger systems. Hollow spheres large enough to be visible in a standard light microscope have been created this way using self-assembling lipids. There are many other examples and this field is rapidly advancing. Biological systems can do most of what molecular nanotechnology strives to accomplish — atomically precise products, active materials, reproduction, etc. However, biological systems are extremely complex and molecular nanotechnology seeks simpler systems to understand, control and manufacture. Also, biological systems usually work at fairly mild temperature and pressure conditions in solution — conditions that are not found in most aerospace environments.

Today, extremely precise atomic and molecular manipulation is common in many laboratories around the world and our abilities are rapidly approaching Feynman's dream. The implications for aerospace development are profound and ubiquitous. A number of applications are mentioned here and a few are described in some detail with references. From this sample of applications it should be clear that although molecular nanotechnology is a long term, high risk project, the payoff is potentially enormous — vastly superior computers, aerospace transportation, sensors and other technologies; technologies that may enable large scale space exploration and colonization. Some of the applications described below have relatively near-term potential and working prototypes may be realized within three to five years. This is certainly not true in other cases. Indeed, many of the possible applications of nanotechnology that we describe here are, at the present time, rather speculative and futuristic. However, each of these ideas have been examined at least cursorily by competent scientists, and as far as we know all of them are within the bounds of known physical laws.

We are not suggesting that their achievement will be easy, cheap or near-term. Some may take decades to realize; some other ideas may be scrapped in the coming years as insuperable barriers are identified. But we feel that they are worth mentioning here as illustrations of the potential future impact of nanotechnology. The applicability of manufacturing at an ever smaller scale is nowhere more self-evident than in computer technology. Indeed, Moore's law (an observation not a physical law) says that computer chip feature size decreases exponentially with time, a trend that predicts atomically precise computers by about 2010-2015. This capability is being approached from many directions. Here we will concentrate on those under development by NASA Ames and her partners. Carbon nanotubes can be viewed as rolled up sheets of graphite from 0.7 to many nanometers in diameter. The smaller tubes are single molecules. Dai placed carbon nanotubes on an SPM tip thus extending our ability to manipulate a single molecule with sub-angstrom accuracy. Not only are the tips atomically precise, but they should have approximately the same chemistry as C_{60}, and thus be functionalizable with a wide variety of molecular fragments. Functionalizing carbon nanotube tips will allow mechanical manipulation of many molecular systems on various surfaces with sub-angstrom accuracy. One particularly intriguing possibility along this line is to utilize a carbon nanotube SPM tip to engrave patterns on a silicon surface. It should be possible to create features a few nanometers across. These would be perhaps 100 times finer than the current state of the art in commercial semiconductor photolithography. Further, in contrast to approaches such as electron microscope lithography for which the speed of operation now appears to be an insuperable obstacle for industrial production, nanotube SPM-based lithography can be accelerated by utilizing an array with thousands of SPM tips simultaneously engraving different parts of a silicon surface.

Also, nanotube SPM lithography could provide a practical means to explore various futuristic electronic device technology ideas, such as quantum cellular automata, which require exceedingly small feature sizes. Needless to say, if these ideas pan out, they could literally revolutionize computer device technology, paving the way for systems that are many times more powerful and more compact than any available today.

For the near term, it should be noted that the semiconductor industry is a major market for SPM products. These are used to examine production equipment. High performance carbon nanotube tips should be of substantial value. NASA Ames is collaborating with Dr. Dai, now at Stanford, to develop these tips. It is possible to store data on long chain molecules (for example, DNA) and it may be possible to read these data with carbon nanotube tipped SPMs. Existing DNA synthesis techniques can be used to write data. If the different DNA base pairs can be distinguished with a carbon nanotube tipped SPM, then the data can be read non-destructively (current techniques allow a destructive read). However, the difference between base pairs is not great. If the base pairs cannot be distinguished, techniques for attaching modified enzymes to specific base pair sequences could be used. Certain enzymes (DNA (cytosine-5) methyltransferases) attach themselves onto a specific sequence of base pairs with a covalent bond.

The enzyme then performs its operation and breaks the bond.

Smith modified the enzyme such that the initial covalent bond was formed but the subsequent operation was disrupted. The result is that DNA synthesized with the target base pair sequences at the desired location can force precise placement of the enzymes. The presence of an enzyme could represent 1 and its absence 0. Enzymes are sufficiently large that distinguishing their presence should be straightforward. If the DNA/enzyme approach proves impossible, a wide variety of other polymer systems could be examined. Bauschlicher computationally studied storing data in a pattern of fluorine and hydrogen atoms on the (111) diamond surface. If write-once data could be stored this way, 10^{15} bytes/cm^2 is theoretically possible. By comparison, the new DVD write-once disks now coming on the market hold about 10^8 bytes/cm^2. Bauschlicher compared the interaction of different probe molecules with a one dimensional model of the diamond surface. This study found some molecules whose interaction energies with H and F are sufficiently different that the force differential should be detectable by an SPM. These studies were extended to include a two dimensional model of the diamond surface and two other systems besides F/H. Other surfaces, such as Si, and other probes, such as those including transition metal atoms, have also been investigated. Among the better probes was C_5H_5N (pyridine).

Quantum calculations suggest that pyridine is stable when attached to C_{60} in the orientation necessary for sensing the difference between hydrogen and fluorine. Half of C_{60} can form the end cap of a (9,0) or (5,5) carbon nanotube, and carbon nanotubes have been attached to an SPM tip. Thus, it might be possible using today's technology to build a system to read the diamond memory surface. Avouris has shown that individual hydrogen atoms can be removed from a silicon surface. If this could be accomplished in a gas that donates fluorine to vacancies on a diamond surface, the data storage system could be built. Thummel computationally investigated methods for adding a fluorine at the radical sites where a hydrogen atom had been removed from a diamond surface. As mentioned before, carbon nanotubes can be described as rolled up sheets of graphite. Different tubes can have different helical windings depending on how the graphite sheet is connected to itself. Theory suggests that single-walled carbon nanotubes can have metallic or semiconductor properties depending on the helical winding of the tube. Chico, Han, Menon, and others have computationally examined the properties of some of hypothetical devices that might be made by connecting tubes with different electrical properties.

Such devices are only few nanometers across — 100 times smaller than current computer chip features. Several authors, including Tour have described methods to produce conjugated macromolecules of precise length and composition. This technique was used to produce molecular electronic devices in mole quantities. The resultant single molecular wires were tested experimentally and found to be conducting. The three and four terminal devices have been examined computationally and look promising. The features of these components are approximately 3 angstroms wide, about 750 times smaller than current silicon technology can produce. Helical logic is a theoretical proposal for a future computing technology using the presence or absence of individual electrons (or holes) to encode 1s and 0s. The electrons are constrained to move along helical paths, driven by a rotating electric field in which the entire circuit is immersed. The electric field remains roughly orthogonal to the major axis of the helix and confines each charge carrier to a fraction of a turn of a single helical loop, moving it like water in an Archimedean screw. Each loop could in principle hold an independent carrier, permitting high information density. One computationally universal logic operation involves two helices, one of which splits into two "descendant" helices. At the point of divergence, differences in the electrostatic potential resulting from the presence or absence of a carrier in the adjacent helix controls the direction taken by a carrier in the splitting helix. The reverse of this sequence can be used to merge two initially distinct helical paths into a single outgoing helical path without forcing a dissipative transition. Because these operations are both logically and thermodynamically reversible, energy dissipation can be reduced to extremely low levels.

It is important to note that this proposal permits a single electron to switch another single electron, and does not require that many electrons be used to switch one electron. The energy dissipated per logic operation can likely be reduced to less than 10^{-27} joules at a temperature of 1 Kelvin and a speed of 10 gigahertz, though further analysis is required to confirm this. Irreversible operations, when required, can be easily implemented and should have a dissipation approaching the fundamental limit of $\ln 2 \times kT$. One study not conducted by Ames or partners is particularly worth mentioning since it places a loose lower bound on the computational capabilities of molecular nanotechnology. Drexler designed a number of computer components using small diamondoid rods with knobs that allow or prevent movement to accomplish computation. While this tiny mechanical Babbage Machine is probably not an optimal computational engine, its calculated performance for a desktop computer is 10^{18} MIPS — about a million times more powerful than the largest supercomputer that exists today. Note that with very fast computation energy use and heat dissipation become a severe problem. One approach to addressing this issue is reversible logic. Drexler proposed a nanotechnology based on diamond and investigated its potential properties. In particular, he examined applications for materials with a strength similar to that of diamond (69 times strength/mass of titanium). This would require a very mature nanotechnology constructing systems by placing atoms on diamond surfaces one or a few at a time in parallel. Assuming diamondoid materials, McKendree predicted the performance of several existing single-stage-to-orbit (SSTO) vehicle designs. The predicted payload to dry mass ratio for these vehicles using titanium as a structural material varied from < 0 (the vehicle won't work) to 36 per cent, i.e., the vehicle weighs substantially more than the payload. With hypothetical diamondoid materials the ratios varied from 243 per cent to 653 per cent, i.e., the payload weighs far more than the vehicle.

Using a very simple cost model ($1000 per vehicle kilogram) sometimes used in the aerospace industry, he estimated the cost per kilogram launched to low-Earth-orbit for diamondoid structured vehicles should be $153-412. This would meet NASA's 2020 launch to orbit cost goals. Estimated costs for titanium structured vehicles varied from $16,000-59,000/kg. Although this cost model is probably adequate for comparison, the absolute costs are suspect. Drexler used a more speculative methodology to estimate that a four passenger SSTO weighing three tons including fuel could be built using a mature nanotechnology. Using McKendree's cost model, such a vehicle would cost about $60,000 to purchase — the cost of today's high-end luxury automobiles. These studies assumed a fairly advanced nanotechnology capable of building diamondoid materials. In the nearer term, it may be possible to develop excellent structural materials using carbon nanotubes. Carbon nanotubes have a Young's modulus of approximately one terapascal— comparable to diamond. Issacs and Pearson proposed a space elevator—a cable extending from the Earth's surface into space with a center of mass at geosynchronous altitude. If such a system could be built, it should be mechanically stable and vehicles could ascend and descend along the cable at almost any reasonable speed using electric power (actually generating power on the way down). The first incredibly difficult problem with building a space elevator is strength of materials. Maximum stress is at geosynchronous altitude so the cable must be thickest there and taper exponentially as it approaches Earth. Any potential material may be characterized by the taper factor—the ratio between the cable's radius at geosynchronous altitude and at the Earth's surface. For steel the taper factor is tens of thousands—clearly impossible. For diamond, the taper factor is 21.9 including a safety factor.

Diamond is, however, brittle. Carbon nanotubes have a strength in tension similar to diamond, but bundles of these nanometer-scale radius tubes shouldn't propagate cracks nearly as well as the diamond tetrahedral lattice. Thus, if the considerable problems of developing a molecular nanotechnology capable of making nearly perfect carbon nanotube systems approximately 70,000 kilometers long can be overcome, the first serious problem of a transportation system capable of truly large scale transfers of mass to orbit can be solved. The next immense problem with space elevators is safety—how to avoid dropping thousands of kilometers of cable on Earth if the cable breaks. Active materials may help by monitoring and repairing small flaws in the cable and/or detecting a major failure and disassembling the cable into small elements. Drexler calculates that lightsails made of 20 nm aluminum in tension should achieve an outward acceleration of ~14 km/s per day at Earth orbit with no payload and minimal structural overhead. For comparison, the delta V from low Earth to geosynchronous orbit is 3.8 km/s. Lightsails generate thrust by reflecting sunlight. Tension is achieved by rotating the sail. The direction of thrust is normal to the sail and away from the Sun. By directing thrust along or against the velocity vector, orbits can be lowered or raised. This form of transportation requires no reaction mass and generates thrust continuously, although the instantaneous acceleration is small so sails cannot operate in an atmosphere and must be large for even moderate payloads. Today, the smallest feature size in production systems is about 250 nanometers — the smallest feature size in computer chips. Since atoms are an angstrom or so across and carbon nanotubes have a diameter as small as 0.7 nanometers, atomically precise molecular machines can be smaller than current MEMS devices by two to three orders of magnitude in each dimension, or six to nine orders of magnitude smaller in volume (and mass). For example, the size of the kinesin motor, which transports material in cells, is 12 nm. Han computationally demonstrated that molecular gears fashioned from single-walled carbon nanotubes with benzyne teeth should operate well at 50-100 gigahertz. These gears are about two nanometers across. Han computationally demonstrated cooling the gears with an inert atmosphere. Srivastava simulated powering the gears using alternating electric fields generated by a single simulated laser. In this case, charges were added to opposite sides of the tube to form a dipole. To make active materials, a material might be filled with nano-scale sensors, computers, and actuators so the material can probe its environment, compute a response, and act.

Although this document is concerned with relatively simple artificial systems, living tissue may be thought of as an active material. Living tissue is filled with protein machines which gives living tissue properties (adaptability, growth, self-repair, etc.) unimaginable in conventional materials. Active materials can theoretically be made entirely of machines. These are sometimes called swarms since they consist of large numbers of identical simple machines that grasp and release each other and exchange power and information to achieve complex goals. Swarms change shape and exert force on their environment under software control. Although some physical prototypes have been built, at least one patent issued, and many simulations run, swarm potential capabilities are not well analyzed or understood. We briefly discuss some concepts here. Michael proposes brick-shaped machines of various sizes that slide past each other to assume a variety of shapes. He has generated a large number of videos showing computer simulations of simple motions. Although his web site contains rather extravagant claims, this work has received a U.K. patent. Yim built a small swarm with macroscopic (size in inches) components called polypod, built a simulator of polypod, and programmed it to move in various ways to study locomotion. There are two brick shaped components in polypod, one of which has two prismatic joints linked by a revolute joint. The second component is a cubic connector with no mechanical motion. Polypod is programmed by tables for each member of the swarm. Each member is programmed to move at various speeds in each degree of freedom for certain amounts of time. The swarm components are implicitly synchronized so there is no clock signal. Hall proposes a swarm with 10 micron dodecahedral components each with 12 arms that can move in and out, rotate a little, and grab and release each other. This concept is called the "utility fog." Hall estimates that the utility fog would have a density of 0.2, tensile strength of 1000 psi in action and 100,000 psi in a passive mode, and have a maximum shear rate of 100 km/second/meter. Bishop proposes a swarm consisting of 100 nanometer brick-shaped components that slide past each other to change shape. Globus proposes a swarm with two kinds of components — edges and nodes. The terms "node" and "edge" are chosen to correspond to those in graph theory.

The roughly spherical nodes are capable of attaching to five edges (for a tetrahedral geometry with one free edge per node) and rotating each edge in pitch and yaw. The rod-like edges are capable of changing length, rotating around their long axis, and attaching/detaching to/from nodes. Component design, power distribution and control software are significant challenges for swarm development. Consider that with 10 micron components a cubic meter of swarm would contain about 10^{15} devices, each with an internal computer communicating with its neighbors to accomplish a global task. NASA's mission is divided into five enterprises: Mission to Planet Earth, Aeronautics, Human Exploration and Development of Space, Space Science, and Space Technology. We will examine some potential nanotechnology applications in each area. The Earth Observing System (EOS) will use satellites and other systems to gather data on the Earth's environment. The EOS data system will need to process and archive >terabyte per day for the indefinite future. Simply storing this quantity of data is a significant challenge—each day's data would fill about 1,000 DVD disks. With projected nanocomputer processing speeds of 10^{18} MIPS, a million calculations on each byte of one day's data would take one second on the desktop. Given a mature nanotechnology, it should be possible to build sensors in balloon-borne systems approximately the size of bacteria.

With replication based manufacturing, these should be quite inexpensive. If the serious communication and control problems can be solved, one can imagine spreading billions of tiny lighter-than-air vehicles into the atmosphere to measure wind currents and atmospheric composition. A similar approach might be taken in the oceans—note that the oceans are full of floating microscopic living organisms that can sense and react to their environment. Smart dust might sense the environment, note the location via a GPS-like system, and store that information until close enough to a data-collection point to transfer the data to the outside world. The strength of materials and computational capabilities previously discussed for space transportation should also allow much more advanced aircraft. Stronger, lighter materials can obviously make aircraft with greater lift and range. More powerful computers are invaluable in the design stage and of great utility in advanced avionics. MEMS technology has been used to replace traditional large control structures on aircraft with large numbers of small MEMS controlled surfaces. This control system was used to operate a model airplane in a windtunnel. Nanotechnology should allow even finer control—finer control than exhibited by birds, some of which can hover in a light breeze with very little wing motion. Nanotechnology should also enable extremely small aircraft.

A reasonably advanced nanotechnology should be able to make simple atomically precise materials under software control. If the control is at the atomic level, then the full range of shapes possible with a given material should be achievable. Aircraft construction requires complex shapes to accommodate aerodynamic requirements. With molecular nanotechnology, strong complex-shaped components might be manufactured by general purpose machines under software control. The aeronautics mission is responsible for launch vehicle development. Payload handling is an important function.

Very efficient payload handling might be accomplished

by a very advanced swarm. The sequence begins by placing each payload on a single large swarm located next to the shuttle orbiter. The swarm forms itself around the payloads and then moves them into the payload bay, arranging the payloads to optimize the center of gravity and other considerations. The swarm holds the payload in place during launch and may even damp out some launch vibrations. On orbit, satellites can be launched from the payload bay by having the swarm give them a gentle push. The swarm can then be left in orbit, perhaps at a space station, and used for orbital operations. This scenario requires a very advanced swarm that can operate in an atmosphere and on orbit in a vacuum. Besides the many and obvious difficulties of developing a swarm for a single environment, this provides additional challenges. Note that a simpler swarm might be used for aircraft payload handling. Aerospace vehicles often require complex checkout procedures to insure safety and reliability. This is particularly true of reusable launch vehicles. A very advanced swarm with some special purpose appendages might be placed on a vehicle. It might then spread out over the vehicle and into all crevices to examine the state of the vehicle in great detail.

Nanotechnology-enabled Earth-to-orbit transportation has the greatest potential to revolutionize human access to space by dropping the current $10,000 per pound cost of launch, but this was discussed above. Other less dramatic technologies include: Space structures with a long design life (such as space station modules) need high-reliability materials that do not degrade. Active materials might help. The machines monitor structural integrity at the sub-micrometer scale. When a portion of the material becomes defective, it could be disassembled and then correctly reassembled. It should be noted that bone works somewhat along these lines. It is constantly being removed and added by specialized cells. To effect timely repairs, space stations require a large store of spare parts and tools that are rarely used. A mature nanotechnology might create a "matter compiler," a machine that converts raw materials into a wide variety of products under software control. Contemporary examples of very limited matter compilers are numerically controlled machines and polypeptide sequencers. With a substantially more capable nanotechnology-based matter compiler, a space station crew could simply make spare parts and tools as needed. The programs could be stored on-board or on the ground. New tools invented on Earth could be transferred as software to the station for manufacture. Once used, unneeded tools and broken parts could be ionized in a solar furnace, transferred using controlled magnetic fields, and the constituent atoms stored for later manufacture into new products. An advanced nanotechnology might be able to build filters that dynamically modify themselves to attract the contaminant molecules detected by the air and water quality sensors. Once attached to the filter, the filter could in principle move the offending molecules to a molecular laboratory for modifications to useful or at least inert products. A swarm might implement such an active filter if it was able to dynamically manufacture proteins that could bind contaminant molecules. The protein and bound contaminant might then be manipulated by the swarm for transportation. With a sufficiently advanced nanotechnology it might even be possible to directly generate food by non-biological means. Then agriculture waste in a self-sufficient space colony could be converted directly to useful nutrition. Making this food attractive will be a major challenge. Sleeping crew members in the shuttle experience considerable pain and sleep disruption when the reaction control system fires and they collide with the cabin walls. If crew members were connected to the walls by a swarm, the swarm could absorb most or all of the force before the crew member struck the wall. The swarm could then gradually return the crew member to center (without the oscillations associated with bungee cords) in preparation for the next firing. For resupply, spacecraft docking is a frequent necessity in space station operations. When two spacecraft are within a few meters of each other, a swarm could extend from each, meet in the middle, and form a stable connection before gradually drawing the spacecraft together. A swarm could support space-suited astronauts in simulated partial-g environments by holding them up appropriately. The swarm moves in response to the astronaut's motion providing the appropriate simulation of partial or 0 gravities.

Tools and other objects are also manipulated by the swarm to simulate non-standard gravity. Active nanotechnology materials (see active materials) might enable construction of a skin-tight space suit covering the entire body except the head (which is in a more conventional helmet). The material senses the astronaut's motions and changes shape to accommodate it. This should eliminate or substantially reduce the limitations current systems place on astronaut range of motion. In situ resource utilization is undoubtedly necessary for large scale colonization of the solar system. Asteroids are particularly promising for orbital use since many are in near Earth orbits. Moving asteroids into low Earth orbit for utilization poses a safety problem should the asteroid get out of control and enter the atmosphere. Very small asteroids can cause significant destruction. The 1908 Tunguska explosion, which was calculated to be a 60 meter diameter stony asteroid, leveled 2,200 km^2 of forest. Hills calculated that 4 meter diameter iron asteroids are near the threshold for ground damage. Both these calculations assumed high collision speeds. At a density of 7.7 g/cm^3 Babadzhanov, a 3 meter diameter asteroid should have a mass of about 110 tons. Rabinowitz estimates that there are about one billion ten meter diameter near Earth asteroids and there should be far more smaller objects. For colonization applications one would ideally provide the same radiation protection available on Earth. Each square meter on Earth is protected by about 10 tons of atmosphere. Therefore, structures orbiting below the van Allen belts would like 10 tons/meter^2 surface area shielding mass. This would dominate the mass requirements of any system and require one small asteroid for each 11 meter^2 of colony exterior surface area. A 10,000 person cylindrical space colony such as Lewis One with a diameter of almost 500 meters and a length of nearly 2000 meters would require a minimum of about 90,000 retrieval missions to provide the shielding mass.

The large number of missions required suggests that a fully automated, replicating nanotechnology may be essential to build large low Earth orbit colonies from small asteroids. A nanotechnology swarm along with an atomically precise lightsail is a promising small asteroid retrieval system. Lightsail propulsion insures that no mass will be lost as reaction mass. The swarm can control the lightsail by shifting mass. When a target asteroid is found, the swarm spreads out over the surface to form a bag. The interface to the sail must be active to account for the rotation of the asteroid—which is unlikely to have an axis-of-rotation in the proper direction to apply thrust for the return to Earth orbit. The active interface is simply swarm elements that transfer between each other to allow the sail to stay in the proper orientation. Of course, there are many other possibilities for nanotechnology based retrieval vehicles. Extraterrestrial materials brought into orbit could be fed into a high-temperature solar furnace and partially ionized. Magnetic fields might then be used to separate the nuclei. These are fed in appropriate quantities to a matter-compiler to build the products desired. Several authors, including Freitas have speculated that a sufficiently advanced nanotechnology could examine and repair cells at the molecular level. Should this capability become available—presumably driven by terrestrial applications—the small size and advanced capabilities of such systems could be of great utility on long duration space flights and on self-sufficient colonies. Self-replicating systems permit efforts of great scope to be pursued economically.

Adjusting the environment on another planet to suit the tastes of humans is one such undertaking. Heating and cooling can be achieved by (among many other methods) using space-based mirrors. Chemical modifications of the planetary surface and atmosphere can be achieved in relatively short periods by the use of self-replicating systems that absorb sunlight and raw materials, and convert them into the desired products. Much as plants changed the environment of the earth to what we see today, so self-replicating molecular manufacturing systems might more rapidly convert the environments of other planets. As interstellar trips might last many years, the ability to conserve supplies by maintaining some crew members in a suspended state would be useful. An extremely advanced nanotechnology might use molecular manipulations of each cell to provide (a) better methods of slowing or suspending the metabolic activity of crew members and (b) better methods of restoring metabolic activity to a normal state when the destination is reached. Molecular manufacturing should enable the creation of very precise mirrors. Unlike lightsail applications, telescope mirrors require a very precise and somewhat complicated shape. A swarm with special purpose appendages capable of bonding to the mirror might be able to achieve and maintain the desired shape. A very advanced nanotechnology would be capable of imaging and then removing the surface atoms of an extra-terrestrial sample. By removing successive surface layers the location of each atom in the sample might be recorded, destroying the sample in the process. This data could then be sent to Earth. Besides requiring a very advanced nanotechnology, there is a more fundamental—but not necessarily fatal—problem: as the outside layer of atoms is removed the next layer may rearrange itself so the sample is not necessarily perfectly recorded. Solar Power—Low Earth orbit spacecraft generally depend on solar cells and batteries for power. For energy collection, molecular manufacturing can be used to make solar photovoltaic cells at least as efficient as those made in the laboratory today. Efficiencies can therefore be > 30 per cent. In space applications, a reflective optical concentrator need consist of little more than a curved aluminum shell < 100 nanometers thick (photovoltaic cells operate with higher efficiency at high optical power densities). A metal fin with a thickness of 100 nanometers and a conduction path length of 100 microns can radiate thermal energy at a power density as high as 1000 W/m^2 with a temperature differential from base to tip of < 1 K.

Accordingly, solar collectors can consist of arrays of photovoltaic cells several microns in thickness and diameter, each at the focus of a mirror of ~100 micron diameter, the back surface of which serves as a ~100 micron diameter radiator. If the mean thickness of this system is ~1 micron, the mass is ~10^{-3} kg/m^2 and the power per unit mass, at Earth's distance from the Sun, where the solar constant is ~1.4 kW/m^2, is > 10^5 W/kg." By comparison, the U.S. built Photovoltaic Panel Module solar cells currently used on the Mir Space Station and planned for use on the International Space Station generate about 118 W/kg. Fuel Cells—A critical component in hydrogen/oxygen fuel cells is the PEM

(Proton Exchange Membrane). This membrane must (a) permit the passage of protons while (b) blocking everything else. Present membranes do a rather poor job. One group at Ames is designing and computationally testing PEMs to study possible energy mechanisms in early life. While these studies are not meant to design optimal membranes for fuel cell use, the basic knowledge and approach may be of value. Another proposal is to design a diamond membrane a few nanometers thick with "proton pores." The pores might be lined with fluorine, oxygen and nitrogen to create a region with a high proton affinity. In addition, a positionally controlled platinum might be held at the mouth of the pore to verify that H_2 can be catalytically split into H+ and e–, and that the barrier for migration of the H+ into the pore is modest in size. Nanotechnology must provide precise control over the manufacturing process of the diamondoid PEM since the pores must be made very precisely. Studies of H_2 absorption and packing in carbon nanotubes and nanoropes are in progress at NASA Ames and elsewhere.

Nanotubes provide large pore sizes and nanoropes have different pore sizes depending on interstitial and other locations. Dillon estimated that the single walled nanotubes in their sample contained 5 to 10 per cent by weight of H_2. The nanotubes were about 0.1 to 0.2 per cent by weight of the total sample. Computational studies at Ames suggest that to store 7-10 per cent H_2 in single walled nanotubes at room temperature the H_2s must be stored inside the tubes, not merely adsorbed on the walls. This work suggests that carbon nanotubes might be developed into an excellent H_2 storage medium within 3-5 years. Calculations with oxygen suggest that a diamondoid sphere ~0.1 microns in diameter should easily hold oxygen at ~1,000 atmospheres. While higher pressures are feasible, they offer declining returns. At higher pressures, the pressure-volume relationship becomes severely non-linear and the density approaches a limiting value. Other gases might also be stored if diamondoid spheres can be built, but the analysis has not been done. High strength light-weight materials will allow greater efficiency of energy storage as angular momentum. Many kinds of ultraminiature electromechanical devices have utility on a miniaturized space craft. It has been shown that manipulating carbon nanotubes changes their electrical properties. This might be exploited to build nanometer scale strain devices. This may be achievable within 3-5 years, and simulations along these lines are in progress. Similar results have been achieved experimentally with C_{60}. The electrical properties of a C_{60} molecule were changed by applying pressure to the molecule with an SPM tip. Smaller, lighter spacecraft are cheaper to launch (current costs are about $10,000/lb) and generally cheaper to build. Diamondoid structural materials can radically reduce structural mass, miniaturized electronics can shrink the avionics and reduce power consumption, and atomically precise materials and components should shrink most other subsystems. Thermal protection is crucial for atmospheric reentry and other tasks. The carbon nanotubes under investigation at NASA Ames and elsewhere may play a significant role.

Most production processes for carbon nanotubes create a tangled mat of nanotubes that has a very low mass-to-volume ratio. Like graphite, the tubes should withstand high temperatures but the tangled mat should prevent them from ablating. This may lead to high temperature applications. Many of the applications discussed here are speculative to say the least. However, they do not appear to violate the laws of physics. Something similar to these applications at these performance levels should be feasible if we can gain complete control of the three-dimensional structure of materials, processes and devices at the atomic scale. How to gain such control is a major, unresolved issue. However, it is clear that computation will play a major role regardless of which approach —positional control with replication, self-assembly, or some other means —is ultimately successful. Computation has already played a major role in many advances in chemistry, SPM manipulation, and biochemistry. As we design and fabricate more complex atomically precise structures, modeling and computer aided design will inevitably play a critical role. Not only is computation critical to all paths to nanotechnology, but for the most part the same or similar computational chemistry software and expertise supports all roads to molecular nanotechnology. Thus, even if NASA's computational molecular nanotechnology efforts should pursue an unproductive path, the expertise and capabilities can be quickly refocused on more promising avenues as they become apparent. As nanotechnology progresses we may expect applications to become feasible at a slowly increasing rate. However, if and when a general purpose programmable assembler/replicator can be built and operated, we may expect an explosion of applications. From this point, building new devices will become a matter of developing the software to instruct the assembler/replicators. Development of a practical swarm is another potential turning point. Once an operational swarm that can grow and divide has been built, a large number of applications become software projects.

Nanotech Products & Applications

There are two general ways available to produce nanomaterials, as shown in the following figure. The first way is to start with a bulk material and then break it into smaller pieces using mechanical, chemical or other form of energy (top-down). An opposite approach is to synthesise the material from atomic or molecular species via chemical reactions, allowing for the precursor particles to grow in size (bottom-up). Both approaches can be done in either gas, liquid, supercritical fluids, solid states, or in vacuum. Most of the manufacturers are interested in the ability to control: a) particle size b) particle shape c) size distribution d) particle composition e) degree of particle agglomeration.

Methods to produce nanoparticles from atoms are chemical processes based on transformations in solution e.g. sol-gel processing, chemical vapour deposition (CVD), plasma or flame spraying synthesis, laser pyrolysis, atomic or molecular condensation. These chemical processes rely on the availability of appropriate "metal-organic" molecules as precursors. Sol-gel processing differs from other chemical processes due to its relatively low processing temperature. This makes the sol-gel process cost-effective and versatile. In spraying processes the flow of reactants (gas, liquid in form of aerosols or mixtures of both) is introduced to high-energy flame produced for example by plasma spraying equipment or carbon dioxide laser. The reactants decompose and particles are formed in a flame by homogeneous nucleation and growth. Rapid cooling results in formation of nanoscale particles. These are chemical processes to materials based on transformations in solution such as sol-gel processing, hydro or solvo thermal syntheses, Metal Organic Decomposition (MOD), or in the vapour phase chemical vapour deposition (CVD). Most chemical routes rely on the availability of appropriate "metal-organic" molecules as precursors. Among the various precursors of metal oxides, namely metal b-diketonates and metal carboxylates, metal alkoxides are the most versatile. They are available for nearly all elements and cost-effective synthesis from cheap feedstock have been developed for some.

Two general ways are available to control the formation

and growth of the nanoparticles. One is called arrested precipitation and depends either on exhaustion of one of the reactants or on the introduction of the chemical that would block the reaction. Another method relies on a physical restriction of the volume available for the growth of the individual nanoparticles by using templates.

The sol gel technique is a long-established industrial process for the generation of colloidal nanoparticles from liquid phase, that has been further developed in last years for the production of advanced nanomaterials and coatings. Sol-gel-processes are well adapted for oxide nanoparticles and composites nanopowders synthesis. The main advantages of sol-gel techniques for the preparation of materials are low temperature of processing, versatility, and flexible rheology allowing easy shaping and embedding. They offer unique opportunities for access to organic-inorganic materials. The most commonly used precursors of oxides are alkoxides, due to their commercial availability and to the high liability of the M-OR bond allowing *facile tailoring in situ* during processing.

Aerosol-based processes are a common method for the industrial production of nanoparticles. Aerosols can be defined as solid or liquid particles in a gas phase, where the particles can range from molecules up to 100 μm in size. Aerosols were used in industrial manufacturing long before the basic science and engineering of the aerosols were understood. For example, carbon black particles used in pigments and reinforced car tires are produced by hydrocarbon combustion; titania pigment for use in paints and plastics is made by oxidation of titanium tetrachloride; fumed silica and titania formed from respective tetrachlorides by flame pyrolysis; optical fibres are manufactured by similar process. Traditionally, spraying is used either to dry wet materials or to deposit coatings. Spraying of the precursor chemicals onto a heated surface or into the hot atmosphere results in precursor pyrolysis and formation of the particles. For example, a room temperature electro-spraying process was developed at Oxford University to produce nanoparticles of compound semiconductors and some metals. In particular, CdS nanoparticles were produced by generating aerosol micro-droplets containing Cd salt in the atmosphere containing hydrogen sulphide. CVD consists in activating a chemical reaction between the substrate surface and a gaseous precursor. Activation can be achieved either with temperature (Thermal CVD) or with a plasma (PECVD: Plasma Enhanced Chemical Vapour Deposition). The main advantage is the nondirective aspect of this technology. Plasma allows to decrease significantly the process temperature compared to the thermal CVD process. CVD is widely used to produce carbon nanotubes.

This method is used mainly for metal containing nanoparticles. A bulk material is heated in vacuum to produce a stream of vaporised and atomised matter, which is directed to a chamber containing either inert or reactive gas atmosphere. Rapid cooling of the metal atoms due to their collision with the gas molecules results in the condensation and formation of nanoparticles. If a reactive gas like oxygen is used then metal oxide nanoparticles are produced.

The theory of gas-phase condensation for the production of metal nanopowders is well known, having been first reported in 1930. Gas-phase condensation uses a vacuum chamber that consists of a heating element, the metal to be made into nano-powder, powder collection equipment and vacuum hardware. The process utilises a gas, which is typically inert, at pressures high enough to promote particle formation, but low enough to allow the production of spherical particles. Metal is introduced onto a heated element and is rapidly melted. The metal is quickly taken to temperatures far above the melting point, but less than the boiling point, so that an adequate vapour pressure is achieved. Gas is continuously introduced into the chamber and removed by the pumps, so the gas flow moves the evaporated metal away from the hot element. As the gas cools the metal vapour, nanometer-sized particles form. These particles are liquid since they are still too hot to be solid. The liquid particles collide and coalesce in a controlled environment so that the particles grow to specification, remaining spherical and with smooth surfaces. As the liquid particles are further cooled under control, they become solid and grow no longer. At this point the nanoparticles are very reactive, so they are coated with a material that prevents further interaction with other particles (agglomeration) or with other materials.

Methods using supercritical fluids are also powerful for the synthesis of nanoparticles. For these methods, the properties of a supercritical fluid (fluid forced into supercritical state by regulating its temperature and its pressure) are used to form nanoparticles by a rapid expansion of a supercritical solution. Supercritical fluid method is currently developed at the pilot scale in a continuous process.

An emerging technology for the manufacture of thin polymer fibers is based on the principle of spinning dilute polymer solutions in a high voltage electric field. Electro spinning is a process by which a suspended drop of polymer is charged with thousands of volts. At a characteristic voltage the droplet forms a Taylor cone, and a fine jet of polymer releases from the surface in response to the tensile forces generated by interaction of an applied electric field, with the electrical charge carried by the jet. This produces a bundle of polymer fibers. The jet can be directed to a grounded surface and collected as a continuous web of fibers ranging in size from a few μm's to less than 100 nm.

Any material containing regular nano-sized pores or voids can be used as a template to form nanoparticles. Examples of such templates include porous alumina, zeolites, di-block co-polymers, dendrimers, proteins and other molecules. The template does not have to be a 3D object. Artificial templates can be created on a plane surface or a gas-liquid interface by forming self-assembled monolayers.

Nanoparticles of a wide range of materials—including a variety of organic and biological compounds, but also inorganic oxides, metals, and semiconductors—can be processed using chemical self-assembly techniques. These techniques exploit selective attachment of molecules to specific surfaces, biomolecular recognition and self-ordering principles (e.g. the preferential docking of DNA strands with complementary base pairs) as well as well-developed chemistry for attaching molecules onto clusters and substrates (e.g. thiol (-SH) end groups) and other techniques like reverse micelle, sonochemical, and photochemical synthesis to realise 1-D, 2-D and 3-D self-assembled nanostructures. The molecular building blocks act as parts of a jigsaw puzzle that join together in a perfect order without an obvious driving force present. Long-term and visionary nanotechnological conceptions, however, go far beyond these first approaches. This applies in particular to the development of biomimetic materials with the ability of self-organisation, self-healing and self-replication by means of molecular nanotechnology. One objective here is the combination of synthetic and biological materials, architectures and systems, respectively, the imitation of biological processes for technological applications. This field of nanobiotechnology is at present still in the state of basic research, but is regarded as one of the most promising research fields for the future.

4

Self-Assembly and Nano-Biotechnology: Crucial Issues

Nanotechnology will be a bottom-up technology, building upward from the molecular scale. It will bring a revolution in human abilities like that brought by agriculture or power machinery. It can even be used to reverse many of the changes brought by agriculture or power machinery. But we humans are huge creations with no direct experience of the molecular world, and this can make nanotechnology hard to visualize, hence hard to understand. Scientists working with moleculas face this problem today. They can often calculate how molecules will behave, but to understand this behavior, they need more than heaps of numbers: they need pictures, movies, and interactive simulations, and so they are producing them at an ever-increasing pace. Molecules are objects that exert forces on one another. If your hands were small enough, you could grab them, squeeze them, and bash them together. Understanding the molecular world is much like understanding any other physical world: it is a matter of understanding size, shape, strength, force, motion, and the like—a matter of understanding the differences between sand, water, and rock, or between steel and soap bubbles. Today's visualization tools give a taste of what will become possible with tomorrow's faster computers and better "virtual realities," simulated environments that let you tour a world that "exists" only as a model inside the computer. Before discussing nanotechnology and how it relates to the technologies of today, let's try to get a more concrete understanding of the molecular world by describing a simulation embedded in a scenario. In this scenario, events and technologies described as dating from 1990 or before are historically accurate; those with later dates are either projections or mere scenario elements. The descriptive details in the simulation are written to fit designs and calculations based on standard scientific data, so the science isn't fiction. We saw Joel Gregory manipulating molecules in the virtual reality of a simulated world using video goggles, tactile gloves, and a supercomputer.

The early twenty-first century should be able to do even better. Imagine, then, that today you were to take a really long nap, oversleep, and wake up decades later in a nanotechnological world. In the twenty-first century, even more than in the twentieth, it's easy to make things work without understanding them, but to a newcomer much of the technology seems like magic, which is dissatisfying. After a few days, you want to understand what nanotechnology is, on a gut level. Back in the late twentieth century, most teaching used dry words and simple pictures, but now—for a topic like this—it's easier to explore a simulated world.

And so you decide to explore a simulation of the molecular world. Looking through the brochure, you read many tedious facts about the simulation: how accurate it is in describing sizes, forces, motions, and the like; how similar it is to working tools used by both engineering students and professionals; how you can buy one for your very own home, and so forth. It explains how you can tour the human body, see state-of-the-art nanotechnology in action, climb a bacterium, etc. For starters, you decide to take an introductory tour: simulations of real twentieth-century objects alongside quaint twentieth-century concepts of nanotechnology. After paying a small fee and memorizing a few key phrases (any variation of "Get me out of here!" will do the most important job), you pull on a powersuit, pocket a Talking Tourguide, step into the simulation chamber, and strap the video goggles over your eyes. Looking through the goggles, you seem to be in a room with a table you know isn't really there and walls that seem too far away to fit in the simulation chamber. But trickery with a treadmill floor makes the walk to the walls seem far enough, and when you walk back and thump the table, it feels solid because the powersuit stops your hand sharply at just the right place. You can even feel the texture of the carvings on the table leg, because the suit's gloves press against your fingertips in the right patterns as you move.

The simulation isn't perfect, but it's easy to ignore the defects. On the table is (or seems to be) an old 1990s silicon computer chip. When you pick it up, as the beginners' instructions suggest, it looks like You feel as though you're falling toward the chip's surface, shrinking rapidly. In a moment, it looks roughly like With your thumb still there holding it. The world grows blurrier, then everything seems to go wrong as you approach the molecular level. First, your vision blurs to uselessness—there is light, but it becomes a featureless fog. Your skin is tickled by small impacts, then battered by what feel like hard-thrown marbles. Your arms and legs feel as though they are caught in turbulence, pulling to and fro, harder and harder. The ground hits your feet, you stumble and stick to the ground like a fly on flypaper, battered so hard that it almost hurts. You asked for realism, and only the built-in safety limits in the suit keep the simulated thermal motions of air molecules and of your own arms from beating you senseless. “Stop!” gives you a rest from the suit’s yanking and thumping, and “Standard settings!” makes the world around you become more reasonable. The simulation changes, introducing the standard cheats. Your simulated eyes are now smaller than a light wave, making focus impossible, but the goggles snap your vision into sharpness and show the atoms around you as small spheres. (Real nanomachines are as blind as you were a moment ago, and can’t cheat.) You are on the surface of the computer chip, between a cell and two blocky nanocomputers like the ones. The simulation keeps bombarding you with air molecules, but the standard settings leave out the sensation of being pelted with marbles. A moment ago you were stuck tight to the ground by molecular stickiness, but the standard settings give your muscles the effective strength of steel—at least in simulation—by making everything around you much softer and weaker. The tourguide says that the only unreal features of the simulation have to do with you—not just your ability to see and to ignore thermal shaking and bombardment, but also your sheer existence at a size too small for anything so complex as a human being.

It also explains why you can see things move, something about slowing down everything around you by a factor of 10 for every factor of 10 enlargement, and by another factor to allow for your being made stronger and hence faster. And so, with your greater strength and some adjustments to make your arms, legs, and torso less sticky, you can stand, see, feel, and take stock of the situation. The ground underfoot, like everything around you, is pebbly with atom-sized bumps the size of your fingertips. Objects look like bunches of transparent grapes or fused marbles in a variety of pretty but imaginary colors. The simulation displays a view of atoms and molecules much like those used by chemists in the 1980s, but with a sharper 3-D image and a better way to move them and to feel the forces they exert. Actually, the whole simulation setup is nothing but an improved version of systems built in the late 1980s—the computer is faster, but it is calculating the same things. The video goggles are better and the whole-body powersuit is a major change, but even in the 1980s there were 3-D displays for molecules and crude devices that gave a sense of touching them.

The gloves on this suit give the sensation of touching whatever the computer simulates. When you run a fingertip over the side of the smaller nanocomputer, it feels odd, hard to describe. It is as if the surface were magnetic–it pulls on your fingertip if you move close enough. But the result isn't a sharp click of contact, because the surface isn't hard like a magnet, but strangely soft. Touching the surface is like touching a film of fog that grades smoothly into foam rubber, then hard rubber, then steel, all within the thickness of a sheet of corrugated cardboard. Moving sideways, your fingertip feels no texture, no friction, just smooth bumps more slippery than oil, and a tendency to get pulled into hollows. Pulling free of the surface takes a firm tug. The simulation makes your atom-sized fingertips feel the same forces that an atom would. It is strange how slippery the surface is—and it can't have been lubricated, since even a single oil molecule would be a lump the size of your thumb. This slipperiness makes it obvious how nano-scale bearings can work, how the parts of molecular machines can slide smoothly. But on top of this, there is a tingling feeling in your fingers, like the sensation of touching a working loudspeaker. When you put your ear against the wall of the nanocomputer, you flinch back: for a moment, you heard a sound like the hiss of a twentieth century television tuned to a channel with no broadcast, with nothing but snow and static—but loud, painfully loud. All the atoms in the surface are vibrating at high frequencies, too fast to see. This is thermal vibration, and it's obvious why it's also called thermal noise. Individual molecules still move too quickly to see. So, to add one more cheat to the simulation, you issue the command "Whoa!", and everything around seems to slow down by a factor of ten. On the surface, you now can see thermal vibrations that had been too quick to follow. All around, air molecules become easier to watch. They whiz about as thick as raindrops in a storm, but they are the size of marbles and bounce in all directions. They're also sticky in a magnetlike way, and some are skidding around on the wall of the nanocomputer. When you grab one, it slips away. Most are like two fused spheres, but you spot one that is perfectly round—it is an argon atom, and these are fairly rare. With a firm grip on all sides to keep it from shooting away like a watermelon seed, you pinch it between your steel-strong fingers. It compresses by about 10 percent before the resistance is more than you can overcome. It springs back perfectly and instantly when you relax, then bounces free of your grip. Atoms have an unfamiliar perfection about them, resilient and unchanging, and they surround you in thick swarms.

At the base of the wall is a churning blob that can only be a droplet of water. Scooping up a handful for a closer look yields a swarm of molecules, hundreds, all tumbling and bumbling over one another, but clinging in a coherent mass. As you watch, though, one breaks free of the liquid and flies off into the freer chaos of the surrounding air: the water is evaporating. Some slide up your arm and lodge in the armpit, but eventually skitter away. Getting rid of all the water molecules takes too much scraping, so you command "Clean me!" to dry off. Beside you, the smaller nanocomputer is a block twice your height, but it's easy to climb up onto it as the tourguide suggests. Gravity is less important on a small-scale: even a fly can defy gravity to walk on a ceiling, and an ant can lift what would be a truck to us. At a simulated size of fifty nanometers, gravity counts for nothing. Materials keep their strength, and are just as hard to bend or break, but the weight of an object becomes negligible. Even without the strength-enhancement that lets you overcome molecular stickiness, you could lift an object with 40 million times your mass–like a person of normal size lifting a box containing a half-dozen fully loaded oil tankers. To simulate this weak gravity, the powersuit cradles your body's weight, making you feel as if you were floating. This is almost like a vacation in an orbital theme park, walking with stickyboots on walls, ceilings, and whatnot, but with no need for antinausea medication. This looks like a cluster of grapes and is about the same size. It even feels a bit like a bunch of grapes, soft and loose. The parts don't fly free like a gas or tumble and wander like a liquid, but they do quiver like gelatin and sometimes flop or twist. It is solid enough, but the folded structure is not as strong as your steel fingers. In the 1990s, people began to build molecular machinery out of proteins, copying biology. It worked, but it's easy to see why they moved on to better materials. From a simulated pocket, you pull out a simulated magnifying glass and look at the simulated protein. This shows a pair of bonded atoms on the surface at 10 times magnification. The atoms are almost transparent, but even a close look doesn't reveal a nucleus inside, because it's too small to see. It would take 1,000 times magnification to be able to see it, even with the head start of being able to see atoms with your naked eye. How could people ever confuse big, plump atoms with tiny specks like nuclei? Remembering how your steel-strong fingers couldn't press more than a fraction of the way toward the nucleus of an argon atom from the air, it's clear why nuclear fusion is so difficult. In fact, the tourguide said that it would take a real-world projectile over a hundred times faster than a high-powered rifle bullet to penetrate into the atomic core and let two nuclei fuse. Try as you might, there just isn't anything you could find in the molecular world that could reach into the middle of an atom to meddle with its nucleus. You can't touch it and you can't see it, so you stop squinting though the magnifying glass. Nuclei just aren't of much interest in nanotechnology. Taking the advice of the tourguide, you grab two molecular knobs on the protein and pull. It resists for a moment, but then a loop comes free, letting other loops flop around more, and the whole structure seems to melt into a writhing coil. After a bit of pulling and wrestling, the protein's structure becomes obvious: It is a long chain–longer than you are tall, if you could get it straight—and each segment of the chain has one of several kinds of knobs sticking off to the side. With the multicolored, glassy-bead portrayal of atoms, the protein chain resembles a flamboyant necklace. This may be decorative, but how does it all go back together? The chain flops and twists and thrashes, and you pull and push and twist, but the original tight, solid packing is lost.

There are more ways to go wrong in folding up the chain than there are in solving Rubik's Cube, and now that the folded structure is gone, it isn't even clear what the result should look like. How did those twentieth-century researchers ever solve the notorious "protein folding problem"? It's a matter of record that they started building protein objects in the late 1980s. This protein molecule won't go back together, so you try to break it. A firm grip and a powerful yank straightens a section a bit, but the chain holds together and snaps back. Though unfolding it was easy, even muscles with the strength of steel—the strength of Superman—can't break the chain itself. Chemical bonds are amazingly strong, so it's time to cheat again. When you say, "Flimsy world–one second!" while pulling, your hands easily move apart, splitting the chain in two before its strength returns to normal. You've forced a chemical change, but there must be easier ways since chemists do their work without tiny superhands. While you compare the broken ends, they thrash around and bump together. The third time this happens, the chain rejoins, as strong as before. This is like having snap-together parts, but the snaps are far stronger than welded steel. Modern assembler chemistry usually uses other approaches, but seeing this happen makes the idea of molecular assembly more understandable: Put the right pieces together in the right positions, and they snap together to make a bigger structure. Remembering the "Whoa!" command, you decide to go back to the properly scaled speed for your size and strength. Saying "Standard settings!," you see the thrashing of the protein chain speed up to hard-to-follow blur. At your feet is a ribbed, ringed cylindrical object about the size of a soup can—not a messy, loosely folded strand like the protein (before it fell apart), but a solid piece of modern nanotechnology. It's a gear like the one. Picking it up, you can immediately feel how different it is from a protein. In the gear, everything is held in place by bonds as strong as those that strung together the beads of the protein chain. It can't unfold, and you'd have to cheat again to break its perfect symmetry. Like those in the wall of the nanocomputer, its solidly attached atoms vibrate only slightly. There's another gear nearby, so you fit them together and make the atomic teeth mesh, with bumps on one fitting into hollows on the other. They stick together, and the soft, slick atomic surfaces let them roll smoothly. Underfoot is the nanocomputer itself, a huge mechanism built in the same rigid style. Climbing down from it, you can see through the transparent layers of the wall to watch the inner works. An electric motor an arm-span wide spins inside, turning a crank that drives a set of oscillating rods, which in turn drive smaller rods. This doesn't look like a computer; it looks more like an engineer's fantasy from the nineteenth century. But then, it is an antique design–the tourguide said that the original proposal was a piece of exploratory engineering dating from the mid-1980s, a mechanical design that was superseded by improved electronic designs before anyone had the tools to build even a prototype. This simulation is based on a version built by a hobbyist many years later. The mechanical nanocomputer may be crude, but it does work, and it's a lot smaller and more efficient than the electronic computers of the early 1990s. It's even somewhat faster. The rods slide back and forth in a blur of motion, blocking and unblocking each other in changing patterns, weaving patterns of logic. This nanocomputer is a stripped-down model with almost no memory, useless by itself. Looking beyond it, you see the other block–which contains a machine powerful enough to compete with most computers built in 1990. This computer is a millionth of a meter on a side, but from where you stand, it looks like a blocky building looming over ten stories tall.

The tourguide says that it contains over 100 billion atoms and stores as much data as a room full of books. You can see some of the storage system inside: row upon row of racks containing spools of molecular tape somewhat like the protein chain, but with simple bumps representing the 1s and 0s of computer data. These nanocomputers seem big and crude, but the ground you're now standing on is also a computer–a single chip roughly as powerful as the smaller, stripped-down nanocomputer at your side. As you gaze out over the chip, you get a better sense for just how crude things were a few decades ago. At your feet, on the smallest scale, the chip is an irregular mess. Although the wall of the nanocomputer is pebbly with atomic-scale bumps, the bumps are as regular as tile. The chip's surface, though, is a jumble of lumps and mounds. This pattern spreads for dozens of paces in all directions, ending in an irregular cliff marking the edge of a single transistor. Beyond, you can see other ridges and plateaus stretching off to the horizon. These form grand, regular patterns, the circuits of the computer. The horizon–the edge of the chip–is so distant that walking there from the center would (as the tourguide warns) take days. And these vast pieces of landscaping were considered twentieth-century miracles of miniaturization? Even back then, research in molecular biology had revealed the existence of smaller, more perfect machines such as the protein molecules in cells. A simulated human cell–put here because earlier visitors wanted to see the size comparisons–its on the chip next to the smaller nanocomputer. The tourguide points out that the simulation cheats a bit at this point, making the cell act as though it were in a watery environment instead of air. The cell dwarfs the nanocomputer, sprawling across the chip surface and rearing into the sky like a small mountain. Walking the nature trail around its edge would lead across many transistor-plateaus and take about an hour.

A glance is enough to show how different it is from a nanocomputer or a gear: it looks organic, it bulges and curves like a blob of liver, but its surface is shaggy with waving molecular chains. Walking up to its edge, you can see that the membrane wrapping the cell is fluid (cell walls are for stiff things like plants), and the membrane molecules are in constant motion. On an impulse, you thrust your arm through the membrane and poke around inside. You can feel many proteins bumping and tumbling around in the cell's interior fluid, and a crisscrossing network of protein cables and beams. Somewhere inside are the molecular machines that made all these proteins, but such bits of machinery are embedded in a roiling, organic mass. When you pull your arm out, the membrane flows closed behind. The fluid, dynamic structure of the cell is largely self healing. That's what let scientists perform experimental surgery on cells with the old, crude tools of the 20th century: they didn't need to stitch up the holes they made when they poked around inside. Even a single human cell is huge and complex.

No real thinking being could be as small as you are in the simulation: A simple computer without any memory is twice your height, and the larger nanocomputer, the size of an apartment complex, is no smarter than one of the submoronic computers of 1990. Not even a bendable finger could be as small as your simulated fingers: in the simulation, your fingers are only one atom wide, leaving no room for the slimmest possible tendon, to say nothing of nerves. For a last look at the organic world, you gaze out past the horizon and see the image of your own, full-sized thumb holding the chip on which you stand.

The bulge of your thumb rises ten times higher than Mount Everest. Above, filling the sky, is a face looming like the Earth seen from orbit, gazing down. It is your own face, with cheeks the size of continents. The eyes are motionless. Thinking of the tourguide's data, you remember: The simulation uses the standard mechanical scaling rules, so being 40 million times smaller has made you 40 million times faster. To let you pull free of surfaces, it increased your strength by more than a factor of 100, which increased your speed by more than a factor of 10. So one second in the ordinary world corresponds to over 400 million here in the simulation. It would take years to see that huge face in the sky complete a single eyeblink. Enough. At the command "Get me out!", the molecular world vanishes, and your feeling of weight returns as the suit goes slack. You strip off the video goggles—and hugely, slowly, blink.

Nanobiotechnology and Nanocomposites: The Relationship

The definition of nanocomposite material has broadened significantly to encompass a large variety of systems such as one-dimensional, two-dimensional, three-dimensional and amorphous materials, made of distinctly dissimilar components and mixed at the nanometer scale. The general class of nanocomposite organic or inorganic materials is a fast growing area of research. Significant effort is focused on the ability to obtain control of the nanoscale structures via innovative synthetic approaches. The properties of nanocomposite materials depend not only on the properties of their individual parents but also on their morphology and interfacial characteristics. This rapidly expanding field is generating many exciting new materials with novel properties. The latter can derive by combining properties from the parent constituents into a single material. There is also the possibility of new properties which are unknown in the parent constituent materials. The inorganic components can be three-dimensional framework systems such as zeolites, two-dimensional layered materials such as clays, metal oxides, metal phosphates, chalcogenides, and even one-dimensional and zero-dimensional materials such as $(Mo_3Se_3^-)_n$ chains and clusters. Experimental work has generally shown that virtually all types and classes of nanocomposite materials lead to new and improved properties when compared to their macrocomposite counterparts. Therefore, nanocomposites promise new applications in many fields such as mechanically reinforced lightweight components, non-linear optics, battery cathodes and ionics, nanowires, sensors and other systems. The general class of organic/inorganic nanocomposites may also be of relevance to issues of bio-ceramics and biomineralization in which *in-situ* growth and polymerization of biopolymer and inorganic matrix is occurring.

Finally, lamellar nanocomposites represent an extreme case of a composite in which interface interactions between the two phases are maximized. Since the remarkable properties of conventional composites are mainly due to interface interactions, the materials dealt with here could provide good model systems in which such interactions can be studied in detail using conventional bulk sample (as opposed to surface) techniques. By judiciously engineering the polymer-host interactions, nanocomposites may be produced with a broad range of properties. Inorganic layered materials exist in great variety. They possess well defined, ordered intralamellar space potentially accessible by foreign species. This ability enables them to act as matrices or hosts for polymers, yielding interesting hybrid nano-composite materials. Lamellar nanocomposites can be divided into two distinct classes, intercalated and exfoliated. In the former, the polymer chains alternate with the inorganic layers in a fixed compositional ratio and have a well defined number of polymer layers in the intralamellar space. In exfoliated nano-composites the number of polymer chains between the layers is almost continuously variable and the layers stand >100 Å apart. The intercalated nanocomposites are also more compound-like because of the fixed polymer/layer ratio, and they are interesting for their electronic and charge transport properties. On the other hand, exfoliated nanocomposites are more interesting for their superior mechanical properties. A subclass of nanocomposites which is the lamellar class of intercalated organic/inorganic nanocomposites and namely those systems that exhibit electronic properties in at least one of the components offers the possibility of obtaining well ordered systems some of which may lead to unusual electrical and mechanical properties. Selected members of this class may be amenable to direct structural characterization by standard crystallographic methods. An important issue in this area is the few structural details that are available, therefore, any system which can be subjected to such analysis is of interest. Nanocomposites also offer the possibility to combine diverse properties which are impossible within a single material, e.g. flexible mechanical properties and superconducting properties. This work is now in its infancy and extensive investigations are underway. Another exciting aspect is the possibility of creating heterostructures composed of different kinds of inorganic layers, which could lead to new multifuntional materials. Several general synthetic routes have been developed for inserting polymer chains into host structures and many novel nanocomposites have been designed.

These include:

- *In-situ* intercalative polymerization (ISIP) of a monomer using the host itself as the oxidant. The rationale behind intercalative polymerization is that host matrices with high electron affinity can oxidatively polymerize appropriate monomers in their interior.
- Monomer intercalation followed by topotactic intralamellar solid state polymerization. This route creates conjugated polymers inside non-oxidizing hosts.
- Direct precipitative encapsulation of polymer chains by colloidally dispersed single layers of a host. This approach gives access to a large variety of nanocomposites with many kinds of polymers and hosts.

Materials with features on the scale of nanometers (10^{-9} meter) often have properties dramatically different from their bulk-scale counterparts. For example, nanocrystalline copper is five times harder than ordinary copper with its micrometer-sized crystalline structure. The development of such materials is currently a research area of great interest. Important among these nanoscale materials are nanocomposites, in which the constituents are mixed on a nanometer-length scale. They often have properties that are superior to conventional microscale composites and can be synthesized using surprisingly simple and inexpensive techniques. A nanoscale dispersion of sheet-like inorganic silicate particles in a polymer matrix, for example, is superior to either constituent in such properties as optical clarity, strength, stiffness, thermal stability, reduced permeability, and flame retardancy. The key to the synthesis of nanocomposites is how the silicates are made to disperse in the polymer. Since the organic and inorganic components are typically immiscible (like water and oil), the silicate surface is modified by attaching surfactant molecules (like adding soap to water). Polymer-based nanocomposites are also being developed for electronics applications such as thin-film capacitors in integrated circuits and solid polymer electrolyes for batteries. Nature also makes fabulous nanocomposites, and scientists are trying to emulate such processes. The abalone shell, for example, has alternating layers of calcium carbonate and a rubbery biopolymer; it is twice as hard as and a thousand times tougher than its components. The study of nanocomposite materials requires a multidisciplinary approach, involving novel chemical techniques and an understanding of physics and surface science. It is a field of broad scientific interest with impressive technological promise. Studies have revealed clearly the property advantages that nanomaterial additives can provide in comparison to both their conventional filler counterparts and base polymer. Properties which have been shown to undergo substantial improvements include:

- Mechanical properties e.g. strength, modulus and dimensional stability.
- Decreased permeability to gases, water and hydrocarbons.
- Thermal stability and heat distortion temperature.
- Flame retardancy and reduced smoke emissions.
- Chemical resistance.
- Surface appearance.
- Electrical conductivity.
- Optical clarity in comparison to conventionally filled polymers.

To date one of the few disadvantages associated with nanoparticle incorporation has concerned toughness and impact performance. Some of the data presented has suggested that nanoclay modification of polymers such as polyamides, could reduce impact performance. Clearly this is an issue which would require consideration for applications where impact loading events are likely. In addition, further research will be necessary to, for example, develop a better under-standing of formulation/structure/property relationships, better routes to platelet exfoliation and dispersion etc. In addition it is important to recognise that nanoparticulate/fibrous loading confers significant property improvements with very low loading levels, traditional microparticle additives requiring much higher loading levels to achieve similar performance. This in turn can result in significant weight reductions (of obvious importance for various military and aerospace applications) for similar performance, greater strength for similar structural dimensions and, for barrier applications, increased barrier performance for similar material thickness. Data provided by Hartmut Fischer of TNO in the Netherlands relating to polyamide – montmorillonite nanocomposites indicates tensile strength improvements of approximately 40 and 20 per cent at temperatures of 23°C and 120°C respectively and modulus improvements of 70 per cent and a very impressive 220 per cent at the same temperatures. In addition Heat Distortion Temperature was shown to increase from 65°C for the unmodified polyamide to 152°C for the nanoclay-modified material, all the above being achieved with just a 5 per cent loading of montmorillonite clay. Similar mechanical property improvements were presented for polymethyl methacrylate—clay hybrids. Further data provided by Akkepeddi of Honeywell relating to polyamide-6 polymers confirms these property trends. In addition, the further benefits of short/long glass fibre incorporation, together with nanoclay incorporation, are clearly revealed.

Nanostructure Synthesis and Biotechnology

There has been steady technological progress in all fields of nanostructure synthesis and assembly, in no small part because of the more general availability of characterization tools having higher spatial, energy, and time resolution to clearly distinguish and trace the process of nanostructure formation. As transmission electron microscopy and X-ray diffraction techniques helped in an earlier period to relate the improved properties of "age-hardened" aluminum alloys to their nanostructure (Koch 1998), today's technological advances in materials characterization are providing new insights into the role of the nanostructure in determining macroscopic properties. The tightly-coupled iteration between characterizing the nanostructure, understanding the relationship between nanostructure and macroscopic material properties, and improved sophistication and control in determining nanostructure size and placement have accelerated the rate of progress and helped to define the critical components of this "new" field of nanostructure science and technology. Tightly focused (1-2 μm), high brightness synchrotron X-ray sources provide detailed structural information on colloids, polymers, alloys, and other material structures, highlighting the inhomogeneities of the material with suitable spatial resolution (Hellemans 1998). Improved characterization - higher spatial resolution - higher sensitivity leads t o motivat es Better control of size and placement Understanding of - structure-property link - influence of interfaces stimulates. Interactive cycle of characterization, understanding and enhanced control in the synthesis and assembly of nanostructures. Another important enabling technology has been the now widely available scanning probe technology, including scanning tunneling microscopy and atomic force microscopy. The power of these techniques has provided impetus for developing even higher performance scanning probe tips, fabricated through microfabrication techniques. Development of different tip structures in various materials has given rise to an entire family of powerful scanning probe techniques that encompass such a wide range of characterization capabilities that one can envision "a laboratory on a tip" (Berger et al. 1996). The development of a tip technology also impacts the synthesis and assembly processes themselves: scanning probe technologies have been used as the basis of materials patterning and processing at nanometer scales (Held et al. 1997; Snow et al. 1997; and Wilder et al. 1997) and have provided information on the mechanical and thermal properties of materials at the nanoscale (Nakabeppu et al. 1995; Tighe et al. 1997; and Zhang et al. 1996). More sophisticated in situ monitoring strategies have provided greater understanding and control in the synthesis of nanostructured building blocks, particularly those formed in vacuum environments. Molecular beam epitaxy (MBE) represents a physical vapor (gas phase) deposition technique where sub-monolayer control can be imposed on the formation of two-dimensional and, more recently, three-dimensional nanostructured materials (Leonard et al. 1993). A great deal of the understanding and control derives from the ability to carry out sensitive monitoring of the growth process in situ: reflection high energy electron diffraction (RHEED) details the nature of the surface and surface bonding, and oscillations of the RHEED intensity provide information on the growth rate (Neave et al. 1983). The improvements brought about by these advances in technology have been substantial, but perhaps of greater importance for this nascent field of nanostructure science and technology has been the development of strategies and technologies that have been formed across the former disciplines. More reliable means of controlling nanostructure size and placement, with an end view of being able to scale up the production of such materials while maintaining the control over the nanostructure, have given impetus to a common search for novel synthesis and assembly strategies.

In that search, it is apparent that the naturally occurring synthesis and assembly of biological materials can provide us with some critical insights. Major efforts in nanoparticle synthesis can be grouped into two broad areas: gas phase synthesis and sol-gel processing. Nanoparticles with diameters ranging from 1 to 10 nm with consistent crystal structure, surface derivatization, and a high degree of monodispersity have been processed by both gas-phase and sol-gel techniques. Typical size variances are about 20%; however, for measurable enhancement of the quantum effect, this must be reduced to less than 5% (Murray et al. 1993). Initial development of new crystalline materials was based on nanoparticles generated by evaporation and condensation (nucleation and growth) in a subatmospheric inert-gas environment (Gleiter 1989; Siegel 1991, 1994).

Various aerosol processing techniques have been reported to improve the production yield of nanoparticles (Uyeda 1991, Friedlander 1998). These include synthesis by combustion flame (Zachariah 1994, Calcote and Keil 1997, Axelbaum 1997, Pratsinis 1997); plasma (Rao et al. 1997); laser ablation (Becker et al. 1997); chemical vapor condensation (Kear et al. 1997); spray pyrolysis (Messing et al. 1994); electrospray (de la Mora et al. 1994); and plasma spray (Berndt et al. 1997). Sol-gel processing is a wet chemical synthesis approach that can be used to generate nanoparticles by gelation, precipitation, and hydrothermal treatment (Kung and Ko 1996). Size distribution of semiconductor, metal, and metal oxide nanoparticles can be manipulated by either dopant introduction (Kyprianidou-Leodidou et al. 1994) or heat treatment (Wang et al. 1997). Better size and stability control of quantum-confined semiconductor nanoparticles can be achieved through the use of inverted micelles (Gacoin 1997), polymer matrix architecture based on block copolymers (Sankaran et al. 1993) or polymer blends (Yuan et al. 1992), porous glasses (Justus et al. 1992), and ex-situ particle-capping techniques (Majetich and Canter 1993; Olshavsky and Allcock 1997). Additional nanoparticle synthesis techniques include sonochemical processing, cavitation processing, microemulsion processing, and highenergy ball milling.

In sonochemistry, an acoustic cavitation process can generate a transient localized hot zone with extremely high temperature gradient and pressure (Suslick et al. 1996). Such sudden changes in temperature and pressure assist the destruction of the sonochemical precursor (e.g., organometallic solution) and the formation of nanoparticles. The technique can be used to produce a large volume of material for industrial applications. In hydrodynamic cavitation, nanoparticles are generated through creation and release of gas bubbles inside the sol-gel solution (Sunstrom et al. 1996). By rapidly pressurizing in a supercritical drying chamber and exposing to cavitational disturbance and high temperature heating, the sol-gel solution is mixed. The erupted hydrodynamic bubbles are responsible for nucleation, growth, and quenching of the nanoparticles. Particle size can be controlled by adjusting the pressure and the solution retention time in the cavitation chamber. Microemulsions have been used for synthesis of metallic (Kishida et al. 1995), semiconductor (Kortan et al. 1990; Pileni et al. 1992), silica (Arriagada and Osseo-Assave 1995), barium sulfate (Hopwood and Mann 1997), magnetic, and superconductor (Pillai et al. 1995) nanoparticles. By controlling the very low interfacial tension (~10-3 mN/m) through the addition of a cosurfactant (e.g., an alcohol of intermediate chain length), these microemulsions are produced spontaneously without the need for significant mechanical agitation. The technique is useful for large-scale production of nanoparticles using relatively simple and inexpensive hardware (Higgins 1997). Finally, high energy ball milling, the only top-down approach for nanoparticle synthesis, has been used for the generation of magnetic (Leslie- Pelecky and Reike 1996), catalytic (Ying and Sun 1997), and structural (Koch 1989) nanoparticles. The technique, which is already a commercial technology, has been considered dirty because of contamination problems from ball-milling processes. However, the availability of tungsten carbide components and the use of inert atmosphere and/or high vacuum processes have reduced impurities to acceptable levels for many industrial applications. Common drawbacks include the low surface area, the highly polydisperse size distributions, and the partially amorphous state of the asprepared powders.

A second category of molecular electronics device that

appears promising for the development of molecular nanoelectronic digital computers is the electric-field controlled molecular switching devices, including molecular quantum-effect devices. These are most closely descended from the solid-state microelectronics and nanoelectronic devices described earlier, and promise to be the fastest and most densely integrated among all of the current alternatives. A group at Purdue University has used self-assembly to fabricate and demonstrate functioning arrays of molecular electronic quantum confinement structures connected by molecular wires. As another example, J.M. Tour proposes embedding a quantum well in a molecular wire by inserting pairs of barrier groups that break the sequence of conjugated porbitals, forming a two-terminal molecular RTD; other structures for three-terminal molecules have been suggested. (In February 1999, Tour reported the demonstration of RTD effects in the manner proposed above; at press time, the paper was still in review [J.M. Tour, personal communication, 1999].) Nondissipative (adiabatic) Thouless electron pumping has also been described at 0.33 K.

Merkle and Drexler consider a hypothetical "helical logic" device in which a single electron can switch another single electron, with 1 or 0 represented as the presence or absence of individual electrons. The electrons are constrained to move along helical paths, driven by a rotating immersive electric field (normal to the helical axis) that confines each charge carrier to a fraction of a turn of a single helical loop, moving it like water in an Archimedean screw. A logic operation involves two helices, one of which splits into two "descendant" helices. At the point of divergence, differences in the electrostatic potential resulting from the presence or absence of a carrier in the adjacent helix control the direction taken by a carrier in the splitting helix; the sequence is reversible, allowing two initially distinct helical paths to merge into a single outgoing helical path without forcing a dissipative transition. Energy dissipation at ~10 GHz is ~10^{-6} zJ/gate-cycle at an operating temperature of 1 K.

Other classes of molecular electronic devices have been studied, including picosecond photoactive or photochromic molecular switching devices using light, and electrochemical molecular devices using electrochemical reactions, to change the shape, orientation, or electron configuration of a molecule in order to switch a current. However, photoactive devices in a dense network would require an NSOM-like emitter to switch individually, since optical photon pathways are not easily confined at length scales very much below optical wavelengths of ~500-1000 nm; electrochemical molecular devices might require immersion in solvent to operate. Fullerenes exhibit a wide range of quantized conductivities that may be useful in nanoelectronic computers. For example, chemisorption alters properties: 3 alkali dopant atoms per C_{60} buckyball gives high conductivity, while none or 6 dopant atoms gives an insulator. Buckytube conductivity also varies with mechanical stress and torsion, tube diameter, and other geometrical parameters. Depending upon the direction of the nanotube chiral vector, carbon nanotubes may be either metallic or semiconducting. Suitably-sized, doped, flexed, stressed, or chiralized fullerenes could be nested to make insulated wires or joined to make nanoelectronic computer components including resistors with quantized "staircase" resistance of $nh/2e^2$, n = 1, 2,... (e is electronic charge, h is Planck's constant), two-terminal devices such as diodes, three- or four-terminal devices such as heterojunction, TUBEFET, or thin-film transistors, and gates that introduce electric fields. A single-molecule transistor using semiconducting buckytubes has been demonstrated experimentally at room temperature.

Carbon nanotubes have been cut to length, size-sorted

by field flow fractionation, their open ends derivatized with thiol groups and then tethered to 10-nm gold particle "junction boxes" which AFM images show can connect together at least two separate tubes. If the tethering process can be spatially controlled using DNA complementarity or by other means, then fullerene nanoelectronic circuit elements could be assembled into three-dimensional CPU-like structures. P. Collins expects an all-carbon 8-bit adder with nanometer dimensions by the year 2002, and J.C. Ellenbogen of MITRE Corp. has designed a molecular half-adder that measures ~10 nm x 10 nm in size. This research area was extremely active in 1998.

Fullerenes may also allow compact memory devices. For example, it has long been speculated that a ~1 nm buckyball encapsulated in a ~1 nm diameter nanotube segment may glide freely back and forth, trapped weakly in each end cap by van der Waals forces. An external voltage could nudge a charged C_{60} molecule to one end of the other, creating a two-state "buckyshuttle RAM" storage device. Such a device could store one bit of information in a ~10 nm^3 volume, a storage density of ~10^8 bits/$micron^3$; the C_{60} thermal velocity of ~100 m/sec at 310 K inside a ~10 nm memory element gives a typical read/write time of ~0.1 nanosec. In early 1999, Kwon and colleagues reported that finely dispersed 4-6 nm diamond powder thermally annealed by heating in a graphite crucible under argon at 1800°C for 1 hour produced oblong multiwall carbon nanostructures which in some cases could move around inside each other, like C_{60} trapped within C_{480}. Kwon noted that there are potential energy minima at either end, so a charged buckyball (e.g., an enclosed ion) could be shuttled with an applied electric field, which could be used for memory storage. Computational simulation of writing a bit showed that the interior buckyball neatly shuttled to the correct end, then gently bounced to rest in ~0.01 nanosec (i.e., ~100 GHz operation) dissipating only ~10 kT in the C_{480}. Kwon proposed a high density memory board using aligned buckyshuttles in a hexagonal lattice with addressing wires above and below, not unlike a ferrite core memory; the memory would likely be nonvolatile at room temperature, with mass production based on self-assembly.

Most contemporary industrial fabrication processes are based on "top-down" technologies, wherein small objects are sawn or machined from larger objects, or small features are imposed on larger objects, in either case by removing unwanted matter. The results of such processes may be small, such as micron-featured integrated circuits, or very large, such as jet aircraft, but in most cases the material is being processed in chunks far larger than molecular scale.

Molecular manufacturing, on the other hand, represents a "bottom-up" technology. Desired products will be built directly by "assembler" machines, molecule by molecule, making larger and larger objects with atomic precision. The results of such processes may also be very small or very large, much as biology builds both micron-sized bacteria and 100-meter tall sequoia trees. However, since assemblers add matter only where it is intended, little need be removed and hence there may be minimal waste during the process. By guiding with precision the assembly of molecules and supra-molecular structures, such a manufacturing system could construct an extraordinarily wide range of products of unprecedented quality and performance.

In 1959, the Nobel physicist Richard P. Feynman observed that: "The principles of physics, as far as I can see, do not speak against the possibility of maneuvering things atom by atom. It is not an attempt to violate any laws; it is something, in principle, that can be done; but, in practice, it has not been done because we are too big . . . Ultimately, we can do chemical synthesis." "A chemist comes to us and says, 'Look, I want a molecule that has the atoms arranged thus and so; make me that molecule. The chemist does a mysterious thing when he wants to make a molecule. He sees that it has got that ring, so he mixes this and that, and he shakes it, and he fiddles around. And, at the end of a difficult process, he usually does succeed in synthesizing what he wants.

"But it is interesting that it would be, in principle, possible (I think) for a physicist to synthesize any chemical substance that the chemist writes down. Give the orders and the physicist synthesizes it. How? Put the atoms down where the chemist says, and so you make the substance. The problems of chemistry and biology can be greatly helped if our ability to see what we are doing, and to do things on an atomic level, is ultimately developed a development which I think cannot be avoided."

Nearly 40 years after Feynman's famous "Plenty of Room at the Bottom" speech, and a decade after Drexler's original proposal for a bottom-up approach to machinebuilding using molecular assemblers, Nobel chemist Richard Smalley also largely agreed that this objective should prove feasible. Noted Smalley: "On a length scale of more than one nanometer, the mechanical robot assembler metaphor envisioned by Drexler almost certainly will work..."

Many skeptical questions arise when one first encounters the ideas of molecular nanotechnology and molecular assemblers. It is useful to keep in mind the proven feasibility of such systems in the biological tradition over billions of years of natural evolution. In one sense, molecular nanotechnology will be a refinement and expansion upon how nature works at the molecular scale. Nature's examples, such as human beings, certainly answer the most basic skeptical questions, such as: Can macroscopic objects be built from molecular scale processes? (Yes, thanks to cellular replication.) Are molecular objects stable? (Of course; the human population alone contains $>10^{29}$ reasonably stable and quite functional ribosomal "nanomachines"). Observes Nobel chemist Jean-Marie Lehn: "The chemist finds illustration, inspiration and stimulation in natural processes, as well as confidence and reassurance since they are proof that such highly complex systems can indeed be achieved on the basis of molecular components." What about quantum effects? The uncertainty principle makes electron positions somewhat fuzzy, but the atom as a whole has a comparatively definite position set by the relatively great mass of the atomic nucleus. The quantum probability function of electrons in atoms tends to drop off exponentially with distance outside the atom, giving atoms a moderately sharp "edge". Mathematically, the positional uncertainty of a single carbon atom of mass $m_C = 2 \times 10^{26}$ kg bound in a single C-C bond of stiffness $k_C = 440$ N/m may be crudely estimated from the classical vibrational frequency $n_C = (k_C/m_C)^{1/2} = 1.5 \times 10^{14}$ Hz. This sets the zero-point vibrational bond energy $E_C = h\, n_C / 2 = 4.9 \times 10^{20}$ J $= k_C\, DX_C^2 / 2$ where $h = 6.63 \times 10^{34}$ J-sec (Planck's constant) and $DX_C \sim 0.015$ nm is the maximum classical amplitude of the bound carbon atom (roughly the same as the 3 dB point for the gaussian wavefunction, notes J. Soreff). Thus DX_C is just ~5% of the typical atomic electron cloud diameter of ~0.3 nm, imposing only a modest additional constraint on the fabrication and stability of nanomechanical structures. (Even in most liquids at their boiling points, each molecule is free to move only ~0.07 nm from its average position.)

How about the effects of Brownian bombardment on nano-machines? Describing a nanomechanical component as a harmonic oscillator embedded in a gas, Drexler notes: "At equilibrium, an impinging gas molecule is as likely to absorb energy as to deliver it, and so molecular bombardment has no net effect on the amplitude of vibration. How a system is coupled to a thermal bath can affect its detailed dynamics, but not the statistical distribution of dynamical quantities." Nanomachines must also obey the laws of thermodynamics. Nanorobots cannot be used as a "Maxwell's demon" — energy must be expended to do useful work; there can be no free lunch.

Will high-energy radiation damage nanomachines? Radiation can break chemical bonds and disrupt molecular machines. The annual failure rate for a properly designed ~1 micron^3 molecular machine (containing many billions of atoms) can be made as low as several percent. This estimate is derived from a consideration of all known thermal, photochemical, radiation, and other damage mechanisms, but the dominant error mechanism that appears difficult to substantially reduce is damage caused by background radiation. Failure rates are not zero, but are nonetheless remarkable by today's standards, especially given error detection and correction at the module level.

During nanodevice fabrication, the problems of crossbonding and reactive intermediates pose additional constraints that must be dealt with using appropriate assembler designs. Specifically, assemblers with rigid positioning components operating in vacuo can restrict reactive intermediates and molecular tool tips to desired locations with a precision of at least ~0.1 nm, or slightly less than an atomic diameter. Within the confines of an evacuated or "eutactic" assembler workspace, the location of every structural atom may be known to within the uncertainty created by thermal noise. Internal structural components may possess relatively high stiffness, so positional uncertainty caused by thermal noise may be small. In the classical analysis, positional variance $DX^2 = kT / k_s$, where $k = 1.381 \times 10^{21}$ J/kelvin (Boltzmann's constant), T is temperature in kelvins, and k_s is component stiffness, typically 10 N/m for nanometer-scale diamondoid components, hence the standard deviation of component position is ~0.02 nm at room temperature, or ~0.01 nm at cryogenic (e.g., LN_2) temperatures. By employing sufficiently stiff mechanisms the position of every structural atom in the system can be known to within a fraction of an atomic diameter with high reliability and without the need for explicit positional sensing. Steric tool hindrance — in molecular manufacturing, the conical volume occupied by a tool tip or manipulator platform — is a technical design problem that may be overcome by using sufficiently stiff tools or multi-tip tools, working in vacuo. Additionally, while some synthetic methods might involve the manipulation of individual atoms by appropriate tools, such as the hydrogen abstraction tool which removes a single selected hydrogen atom from a diamondoid surface, other reactions will involve small clusters of atoms or larger molecular components.

What about friction and wear among nanomechanical components? Properly designed molecular machines lack wear mechanisms, although other damage mechanisms remain. Dry bearings with atomically precise surfaces can have negligible static friction and wear, and very low dynamic friction or drag. Nanomechanical components are better viewed as moving smoothly in a force field than as sliding subject to friction. For example, total drag power dissipated by an isolated molecular sleeve bearing of radius 2 nm spinning at 1 MHz is dominated by band-stiffness scattering, amounting to 0.000004 picowatts (pW), very small compared to the 1-1000 pW power draw that will be typical of micron-size medical nanorobots. The strong precise surfaces of nanomachines experience no change during a typical operational cycle, hence zero wear. Within the single-point failure model, the first step in a wear process (e.g., a dislocated atom) is regarded as fatal, hence cumulative wear plays no role in determining device lifetimes. In eutactic nanomachines, contaminants of all kinds are rigorously excluded. The equivalent of wear particles cannot appear until the device has failed. On this scale, repulsive fields provide "lubrication" and an oil molecule would be a contaminant object, not a lubricant.

The general conclusion is that molecular nanotechnology violates no physical laws. In 1998, it was generally accepted that molecular nanotechnology will be developed, although there remains some disagreement about how long it will take. To progress from today's limited capabilities to the complex nanorobots described later in this trilogy will clearly require a great deal of research and development effort. Still, the proper question is no longer "if" but "when." This Chapter presents a few of the many possible technical paths leading to a working molecular assembler that were being discussed in 1998. The reader is strongly urged to consult the most current literature for the latest results. The conventional top-down approaches to nanotechnology. Several possible direct bottom-up pathways to molecular manufacturing including biotechnology, supramolecular chemistry, and scanning probes. Gimzewski and Joachim note that "bottom-up approaches to nanofabrication may one day compete with conventional top-down approaches in providing nanotechnologies for the next millenium. Top-down refers to increasing miniaturization through extension of existing microfabrication schemes, [whereas] the bottom-up scenario is one of ever-increasing complexity at the molecular level while maintaining control on an atom-by-atom basis". Surveys molecular mechanical components and design work on molecular assemblers and molecular manufacturing systems from the perspective of 1998.

One of the most challenging problems in synthesis is the controlled generation of monodispersed nanoparticles with size variance so small that size selection by centrifugal precipitation or mobility classification is not necessary. Among all the synthesis techniques discussed above, gas-phase synthesis is one of the best techniques with respect to size monodispersity, typically achieved by using a combination of rigorous control of nucleationcondensation growth and avoidance of coagulation by diffusion and turbulence as well as by the effective collection of nanoparticles and their handling afterwards. The stability of the collected nanoparticle powders against agglomeration, sintering, and compositional changes can be ensured by collecting the nanoparticles in liquid suspension. For semiconducting particles, stabilization of the liquid suspension has been demonstrated by the addition of polar solvent (Murray et al. 1993); surfactant molecules have been used to stabilize the liquid suspension of metallic nanoparticles.

Alternatively, inert silica encapsulation of nanoparticles by gas-phase reaction and by oxidation in colloidal solution has been shown to be effective for metallic nanoparticles (Andres et al. 1998). New approaches need to be developed for the generation of monodisperse nanoparticles that do not require the use of a size classification procedure. An example of this is a process developed in Japan where very monodispersed gold colloidal nanoparticles with diameters of about 1 nm have been prepared by reduction of metallic salt with UV irradiation in the presence of dendrimers (Esumi et al. 1998). Poly(amidoamine) dendrimers with surface amino groups of higher generations have spherical 3-D structures, which may have an effective protective action for the formation of gold nanoparticles. Although the specific role of dendrimers for the formation of monodispersed nanoparticles has yet to be defined, good monodispersity is thought to come from the complex reaction accompanying the decomposition of dendrimers, which eventually leads to the conversion of solution ions to gold nanoparticles.

Scaleup production is of great interest for nanoparticle synthesis. High energy ball milling, already a commercial high-volume process, as mentioned above, has been instrumental in generating nanoparticles for the preparation of magnetic, structural, and catalytic materials. However, the process produces polydispersed amorphous powder, which requires subsequent partial recrystallization before the powder is consolidated into nanostructured materials. Although gas-phase synthesis is generally a low production rate process (typically in the 100 milligrams per hour range) in research laboratories, higher rates of production (about 20 grams per hour) are being demonstrated at Ångström Laboratory at Uppsala University in Sweden. Even higher production rates (about 1 kg per hour) are now being achieved commercially. For sol-gel processing, the development of continuous processing techniques based on present knowledge of batch processing has yet to be addressed for economical scaleup production of nanoparticles. Other related sol-gel issues concern the cost of precursors and the recycling of solvent. Overall, sol-gel processing is attractive for commercial scale-up production.

A recent paradigm shift envisioned for optoelectronics and computational devices involves the assembly of molecular or quantum wires (Chidsey and Murray 1986). Large polymeric molecules have been used as nano building blocks for nanoporous molecular sieves, biocompatible materials, optical switching, data processing, and other nonlinear optical components. Chain aggregates of nanoparticles can be considered as polymer-like units with their primary particles composed of a few hundred to a few thousand molecules. Thus, these chain aggregates can be considered "heavy" quantum wires. In fact, nanoparticle chain aggregates have been studied extensively as magnetic materials (Zhang and Manthiram 1997), as reinforced elastomers (Pu et al. 1997), and as additives in concrete (Sabir 1997). These aggregates have been shown to have chemical and mechanical properties different from those of individual primary particles (Friedlander et al. 1998). Depending on the particle size and its compositional material, the bonding force responsible for holding the aggregates together varies from weak van der Waals force for micrometer particles to strong chemical bonds for nanometer particles to very strong magnetic dipolar bonds for nanosized magnetic particles. The mechanical, optical, and electronic transport properties can be varied by controlling the diameter and the monodispersity of the primary particles, the crystalline structure and morphology, aggregate length, interfacial properties, and material purity. These chain aggregates can be formed by allowing agglomeration of nanoparticles generated by any of the synthesis techniques discussed above, with the exception of high energy ball milling, which generates particles with low surface area and high anisotropic morphologies, both of which are detrimental for the formation of chain aggregates. Depending on the magnetic and electric charging properties of the nanoparticles, an external applied magnetic or electric field can be used to control the fractal dimension of aggregates. For optical applications of chain aggregates, lower fractal dimensions (i.e., relatively straight chain aggregates with few branches) are desirable.

Recent advances in the fabrication of nanometer fibers or tubes offer another form of building blocks for nanostructured materials. An effective way to generate nanometer fibers (or tubes) is based on the use of membranetemplate techniques (Martin 1994). Membranes, with nanochannels generated by fission-fragment tracks or by electrochemical etching of aluminum metal, are used as templates for either chemical or electrochemical deposition of conductive polymers (Pathasarathy and Martin 1994), metal (van de Zande et al. 1997), semiconductor (Klein et al. 1993), and other materials for the generation of nanofibers or tubes. Since the nanochannels on membranes are very uniform in size, the diameter and the aspect ratio of the nanofibers (or tubes) synthesized by the membranetemplate technique can be precisely controlled. This greatly facilitates the interpretation of optical data and the processing of these fibers (or tubes) into 2-D nanostructured materials (de Heer et al. 1995). Single-crystal semiconductor nanofibers can also be grown catalytically by metalorganic vapor phase epitaxy and laser ablation vapor-liquid-solid techniques (Hiruma et al. 1995; Morales and Lieber 1998). The synthesis of these onedimensional structures with diameters in the range of 3 to 15 nm holds considerable technological promise for optoelectronic device applications, such as the p-n junctions for light emission at Hitachi Central Research Laboratory in Japan. The advent of carbon-based nanotubes has created yet another way to fabricate nanometer fibers and tubes. These nanotubes have been used as templates for the fabrication of carbide and oxide nanotubes (Dai et al. 1995; Kasuga et al. 1998). Synthesis of nanotubes based on BN, BC3 and BC2N have also been reported (Chopra et al. 1995; Miyamoto et al. 1994). These nanotubes potentially possess large third-order optical non-linearity and other unusual properties (Xie and Jiang 1998). Metallic nanofibers synthesized by carbon-nanotube-template techniques are useful in the design of infrared absorption materials. The carbon nanotubes can now be catalytically produced in large quantities and have been used for reinforcement of nanostructural composite materials and concrete (Peigney et al. 1997). The elegant complexity of biological materials represents the achievement of structural order over many length scales, with the full structure developed from the "nested levels of structural hierarchy" (Aksay et al. 1992), in which self-assembled organic materials can form templates or scaffolding for inorganic components. These notions of a multilevel material structure with strong interactions among levels and an interplay of perfection and imperfection forming the final material was discussed earlier by Cyril Stanley Smith (1981). Along with characteristic length scales, there are characteristic relaxation times of the material, bringing in the consideration of the temporal stability of the structured materials (Zener 1948). A more detailed discussion of the role of biological materials as both paradigm and tool for the fabrication of nanostructured materials is given. It is interesting how many of the synthesis and assembly approaches seek out and adapt two key features of biogenic fabrication: that of "self-assembly," and the use of natural nanoscaled templates.

Self-Assembly

"Self-assembly" is a term that by now figures prominently in the literature of nanostructured materials and nanofabrication. The term therefore carries a variety of implicit or explicit meanings, and we cite the definition given by Kuhn and Ulman: "Self-assembly is a process in which supermolecular hierarchical organization is established in a complex system of interlocking components." The mechanisms that produce the hierarchical organization are determined by competing molecular interactions (e.g., interactions between hydrophobic versus hydrophilic components, van der Waals, Coulombic, or hydrogen bonding), resulting in particular microphase separation or surface segregation of the component materials. Thus, the use of a hierarchy of bond strengths and/or chemical specificity can produce a hierarchy of lengths in the final nanostructured material (Muthukumar et al. 1998; Stupp and Braun 1998). As one example of such self-assembled or self-organized materials, McGehee et al. (1994) have mixed silica precursors with surfactants that have self-ordered to form various surfactant-water liquid crystals, producing various structures built from walls of amorphous silica, organized about a repetitive arrangement of pores up to a hundred angstroms in diameter. In the "natural" formation of inorganic nanostructures, the addition of organic molecules can strongly influence the resulting structure of inorganic components. Such strategies have been adopted in synthetic formation of nanostructures, such as in the formation of networks of gold clusters (Andres et al. 1998). Gold clusters 3.7 nm in diameter, formed in the vapor phase, are encapsulated in organic surfactants, such as dodecanethiol, forming a colloidal suspension. The surfactants prevent the agglomeration of the gold clusters. Addition of small amount of dithiol precipitates out a 3-D cluster network, which can in turn be deposited onto another solid substrate. A transmission electron microscope (TEM) image of a cluster array spin-cast onto MoS2. The methodology of self-assembly has even been extended to physical vapor deposition processes where it would seem more difficult to control the nucleation and growth of three-dimensional nanostructures. Utilizing the strain inherent in the epitaxial growth of lattice-mismatched materials, and the expected strain-induced transition from two-dimensional (layered) to three-dimensional (islanded) growth, together with careful monitoring of the growth process through RHEED analysis, researchers have been able to form arrays of semiconductor quantum dots, ~ 200-300 Å in diameter, ~ 1011 cm-2 in density, and with a size variation of about ±7% (Leonard et al. 1993). An example of such "self-assembled" semiconductor quantum dots.

The achievement of arrays of several billions of quantum dots of these dimensions with such a size variation is beyond the capability of standard high resolution lithographic and pattern transfer processes. Moreover, the controlled formation of critical surfaces and interfaces without the intercession of ion-assisted processing that can introduce potential defects into the materials has produced a rich source of optically and electronically efficient quantum structures. A number of researchers have already incorporated such self-assembled dots into laser structures (Bimberg et al. 1998). Chemical specificity may provide the most robust means of ensuring control of size and placement of nanostructured building blocks, and recent work in the synthesis of compound semiconductor quantum dots from chemical precursors have provided even tighter distributions of size variation (±5%) than those shown in the strain-induced self-assembled dots (Katari et al. 1994; Murray et al. 1993).

Natural Templates or "Scaffolds"

Biogenic systems employ natural templates or scaffolding in the construction of nanoscaled materials, where the templates and scaffolding can help set the proper, critical dimensions that are characteristic of the final material. The templates can be formed "artificially," through the use of lithographically defined patterning and processing of the substrate material in order to achieve selective nucleation and growth of the nanostructured materials. However, there are also a host of "natural" templating materials that researchers can avail themselves of. The zeolites have for a long time been utilized as the basis of high surface area materials that enable catalysis. Recent work has taken these natural cage structures and formed large cage structures with varying pore sizes (Estermann et al. 1991; Ying and Sun 1997; Bu et al. 1998). When zeolite cages are "loaded" with various materials, the controlled proximity enforced by the cage structure can result in various magnetically active or nonlinear optical materials. More recently, the C-based fullerenes and C-nanotubes have provided a tremendously rich basis for nanostructure research; the impact of such materials is further described. The by now "easy" formation of such structures and ready availability of such materials provide a rich source of nanostructures with nearly immediate applications for electronic devices, storage, and enhancement of structural materials. Further, these materials can serve as the nanostructured precursors of nanostructure synthesis approaches (such as the starting material for cluster or aerosol deposition), or can provide an avenue for the synthesis of templates of varying sizes. Such templates can be electronically "doped" and filled with other materials. Recent work reported the synthesis of "nanorods" of GaN through confined reactions in carbon nanotubes (Han et al. 1998).

Opportunities and Challenges

The broad explorations of this WTEC study have shown a great diversity of impressive work on nanostructured materials. One could argue that many aspects of the work on nanostructured materials have been long-established efforts, with well-developed techniques that have been brought forth into the manufacturing arena. How do we then explain the current excitement and interest in nanostructured materials as a promising new endeavor? Part of the answer lies in the recognition of common scientific issues and common enabling technologies that link this group of researchers together. Ready availability of ever more sophisticated characterization methods that allow us to visualize and probe materials at the nanoscale has accelerated the pace of activities in this field; at the same time, recognition of the common critical issues of control over nanostructure size and placement motivates sharing of solutions over the boundaries of conventional disciplines. Thus, some of the most exciting findings of this study manifest the crossfertilization of techniques and ideas; for example, aerosol particles are being integrated with more traditionally fabricated electronic nanostructures, with placement achieved through the manipulation of STM tips, in order to explore ideas of electronic device enhancement at the nanostructure level (Junno et al. 1998). The ordered assembly of diblock copolymers define three-dimensional nanostructures, and those structures are transferred into electronic (semiconductor) substrates through high resolution pattern transfer processes (Möller 1998). Recognizing the need for effective utilization of ideas across disciplines brings the responsibility of establishing the educational infrastructure that will adequately train young scientists to more fully develop the concepts and applications of nanostructured materials in the future.

Such an infrastructure will require researchers and educators who are familiar with the properties of a broad range of materials, including polymers, biomolecular materials, metals, ceramics, and semiconductors. It will benefit from an informed perspective on critical applications, and must provide access to a wide range of synthesis and assembly techniques and characterization methods that currently reside separately in the disciplines of physics, chemistry, electrical engineering, biology, etc. Finally, a critical enabler for the future of this field and for its educational infrastructure is further development of computational tools that encompass the full range of atomistic calculations to macroscopic materials properties. This will require a "systems approach" (Olson 1997) that will span a variety of different computational models, addressing the different critical length scales, starting with solutions of the Schrodinger and Poisson equations, solving interatomic forces, and scaling all the way up to simulations of macroscopic properties and behavior (Goddard 1998). Increased appreciation of and access to the diverse means of nanoparticle synthesis and assembly that have been developed within many different disciplines, and a common development of enabling tools and technologies, will enhance the pace of accomplishments in this new area of nanoscale synthesis and assembly.

Enabling Technologies

There has been steady technological progress in all fields of nanostructure synthesis and assembly, in no small part because of the more general availability of characterization tools having higher spatial, energy, and time resolution to clearly distinguish and trace the process of nanostructure formation. As transmission electron microscopy and X-ray diffraction techniques helped in an earlier period to relate the improved properties of "age-hardened" aluminum alloys to their nanostructure (Koch 1998), today's technological advances in materials characterization are providing new insights into the role of the nanostructure in determining macroscopic properties. The tightly-coupled iteration between characterizing the nanostructure, understanding the relationship between nanostructure and macroscopic material properties, and improved sophistication and control in determining nanostructure size and placement have accelerated the rate of progress and helped to define the critical components of this "new" field of nanostructure science and technology. Tightly focused (1-2 μm), high brightness synchrotron X-ray sources provide detailed structural information on colloids, polymers, alloys, and other material structures, highlighting the inhomogeneities of the material with suitable spatial resolution (Hellemans 1998). Improved characterization - higher spatial resolution - higher sensitivity leads t o motivat es Better control of size and placement Understanding of - structure-property link - influence of interfaces stimulates. Interactive cycle of characterization, understanding and enhanced control in the synthesis and assembly of nanostructures. Another important enabling technology has been the now widely available scanning probe technology, including scanning tunneling microscopy and atomic force microscopy. The power of these techniques has provided impetus for developing even higher performance scanning probe tips, fabricated through microfabrication techniques. Development of different tip structures in various materials has given rise to an entire family of powerful scanning probe techniques that encompass such a wide range of characterization capabilities that one can envision "a laboratory on a tip" (Berger et al. 1996). The development of a tip technology also impacts the synthesis and assembly processes themselves: scanning probe technologies have been used as the basis of materials patterning and processing at nanometer scales (Held et al. 1997; Snow et al. 1997; and Wilder et al. 1997) and have provided information on the mechanical and thermal properties of materials at the nanoscale (Nakabeppu et al. 1995; Tighe et al. 1997; and Zhang et al. 1996).

More sophisticated in situ monitoring strategies have provided greater understanding and control in the synthesis of nanostructured building blocks, particularly those formed in vacuum environments. Molecular beam epitaxy (MBE) represents a physical vapor (gas phase) deposition technique where sub-monolayer control can be imposed on the formation of two-dimensional and, more recently, three-dimensional nanostructured materials (Leonard et al. 1993). A great deal of the understanding and control derives from the ability to carry out sensitive monitoring of the growth process in situ: reflection high energy electron diffraction (RHEED) details the nature of the surface and surface bonding, and oscillations of the RHEED intensity provide information on the growth rate (Neave et al. 1983). The improvements brought about by these advances in technology have been substantial, but perhaps of greater importance for this nascent field of nanostructure science and technology has been the development of strategies and technologies that have been formed across the former disciplines.

More reliable means of controlling nanostructure size and placement, with an end view of being able to scale up the production of such materials while maintaining the control over the nanostructure, have given impetus to a common search for novel synthesis and assembly strategies. In that search, it is apparent that the naturally occurring synthesis and assembly of biological materials can provide us with some critical insights. Major efforts in nanoparticle synthesis can be grouped into two broad areas: gas phase synthesis and sol-gel processing. Nanoparticles with diameters ranging from 1 to 10 nm with consistent crystal structure, surface derivatization, and a high degree of monodispersity have been processed by both gas-phase and sol-gel techniques. Typical size variances are about 20%; however, for measurable enhancement of the quantum effect, this must be reduced to less than 5% (Murray et al. 1993). Initial development of new crystalline materials was based on nanoparticles generated by evaporation and condensation (nucleation and growth) in a subatmospheric inert-gas environment (Gleiter 1989; Siegel 1991, 1994). Various aerosol processing techniques have been reported to improve the production yield of nanoparticles (Uyeda 1991, Friedlander 1998). These include synthesis by combustion flame (Zachariah 1994, Calcote and Keil 1997, Axelbaum 1997, Pratsinis 1997); plasma (Rao et al. 1997); laser ablation (Becker et al. 1997); chemical vapor condensation (Kear et al. 1997); spray pyrolysis (Messing et al. 1994); electrospray (de la Mora et al. 1994); and plasma spray (Berndt et al. 1997). Sol-gel processing is a wet chemical synthesis approach that can be used to generate nanoparticles by gelation, precipitation, and hydrothermal treatment (Kung and Ko 1996). Size distribution of semiconductor, metal, and metal oxide nanoparticles can be manipulated by either dopant introduction (Kyprianidou-Leodidou et al. 1994) or heat treatment (Wang et al. 1997). Better size and stability control of quantum-confined semiconductor nanoparticles can be achieved through the use of inverted micelles (Gacoin 1997), polymer matrix architecture based on block copolymers (Sankaran et al. 1993) or polymer blends (Yuan et al. 1992), porous glasses (Justus et al. 1992), and ex-situ particle-capping techniques (Majetich and Canter 1993; Olshavsky and Allcock 1997). Additional nanoparticle synthesis techniques include sonochemical processing, cavitation processing, microemulsion processing, and highenergy ball milling. In sonochemistry, an acoustic cavitation process can generate a transient localized hot zone with extremely high temperature gradient and pressure (Suslick et al. 1996). Such sudden changes in temperature and pressure assist the destruction of the sonochemical precursor (e.g., organometallic solution) and the formation of nanoparticles.

The technique can be used to produce a large volume of material for industrial applications. In hydrodynamic cavitation, nanoparticles are generated through creation and release of gas bubbles inside the sol-gel solution (Sunstrom et al. 1996). By rapidly pressurizing in a supercritical drying chamber and exposing to cavitational disturbance and high temperature heating, the sol-gel solution is mixed. The erupted hydrodynamic bubbles are responsible for nucleation, growth, and quenching of the nanoparticles. Particle size can be controlled by adjusting the pressure and the solution retention time in the cavitation chamber. Microemulsions have been used for synthesis of metallic (Kishida et al. 1995), semiconductor (Kortan et al. 1990; Pileni et al. 1992), silica (Arriagada and Osseo-Assave 1995), barium sulfate (Hopwood and Mann 1997), magnetic, and superconductor (Pillai et al. 1995) nanoparticles. By controlling the very low interfacial tension (~10-3 mN/m) through the addition of a cosurfactant (e.g., an alcohol of intermediate chain length), these microemulsions are produced spontaneously without the need for significant mechanical agitation. The technique is useful for large-scale production of nanoparticles using relatively simple and inexpensive hardware (Higgins 1997). Finally, high energy ball milling, the only top-down approach for nanoparticle synthesis, has been used for the generation of magnetic (Leslie- Pelecky and Reike 1996), catalytic (Ying and Sun 1997), and structural (Koch 1989) nanoparticles. The technique, which is already a commercial technology, has been considered dirty because of contamination problems from ball-milling processes. However, the availability of tungsten carbide components and the use of inert atmosphere and/or high vacuum processes have reduced impurities to acceptable levels for many industrial applications. Common drawbacks include the low surface area, the highly polydisperse size distributions, and the partially amorphous state of the asprepared powders.

Opportunities in Self-assembly

In self-assembly large molecular structures are obtained from the organization of a large number of molecules or atoms into a given shape, typically through specific interactions of the molecules among themselves and with a template. The interaction of the different bonding mechanisms is an area of strong fundamental research interest. Only two areas will be highlighted here: zeolites and carbon materials. Both of these materials exhibit characteristics of self-assembly, namely novel and reproducible structures that can be fabricated in industrially significant quantities.

Zeolitic Materials

Aluminosilicates (e.g., zeolites) are crystalline porous nanostructures with long range crystalline order with pore sizes which can be varied from about 4 Å to 15 Å in conventional zeolites. A 3-dimensional (e.g., MFI) zeolite cage structure together with a depiction of the straight and ziz-zag channels and a 2-dimensional zeolite with channels only in 2 directions. The vertices in the stick drawing denote position of the O atoms in the crystalline lattice. This particular zeolite has 10 atoms in the zeolite "window." The size of the window is determined by the number of oxygens in the ring. Typical zeolite structures together depicting the positions of the O atoms and two different zeolitic structures one (lower left) with a three dimensional structure and (lower right) a zeolite with a two dimensional channel structure. For example normal hexane with a kinetic molecular diameter of about 5.1 Å can pass through a 10 ring or larger, whereas cyclohexane with a kinetic molecular diameter of 6.9 Å would be hard pressed to pass through a 10 ring. Thus all other things being equal, a 10-ring zeolite could be used to separate mixtures of normal hexane and cyclohexane. It is this property together with the ability to chemically modify the acidity of zeolitic materials that makes them extremely valuable as selective sorbants, as membranes and for use in selective catalytic reactions. In 1992, a new family of aluminosilicates (M41S) with pores sizes between 20 and 100 Å in diameter were reported by Mobil researchers (Beck et al. 1992; Kresge et al. 1992). One of particular interest is MCM-41, which consists of hexagonal arrays of uniform 2 to 10 nanometer-sized cylindrical pores. Not only can such materials be synthesized, but novel structures such as "tubules-within-a-tubule" have been fabricated as mesoporous molecular sieves in MCM-41 (Lin and Mou 1996). Of particular interest is the possibility of expanding the so-called "liquid crystal templating" mechanism (Chen et al. 1993) to non-aluminum dopants within the silicate MCM-41 framework (Tanev et al. 1994) and to derive nonsiliceous MCM-41 type of materials (Braun et al. 1996). Another approach to synthesizing large pore and large single crystals of zeolytic materials is being pioneered by Geoffrey Ozin and his group at the University of Toronto, who have demonstrated that crystals as large as 5 mm can be synthesized (Kupperman et al. 1993). The ability to synthesize such large crystals has important implications for discovery of new sensors (selective chemical adsorbants) and membrane devices (selective transport of molecular species), since large single crystals can now be available to the laboratory researcher to carry out fundamental studies of adsorption and diffusion properties with such materials. These materials are expected to create new opportunities for applications in the fields of separations science, for use directly as molecular sieves or as new molecular sieving sorbant materials; in catalysis, as heterogeneous catalysts; and as supports for other catalytic materials as well as other novel applications (Bowes et al. 1996; Brinker 1996; Sayari 1996). The ability to synthesize zeolitic materials of precise pore size in the range between 4 and 100 Å continues to expand the possibilities for research and technological innovation in the catalytic, separations, and sorption technologies (Ruthven et al. 1994; Karger and Ruthven 1992).

Carbon Materials

The carbon-based materials of interest from a molecular self-assembly point of view include fullerenes and their relatives, including endohedral fullerenes and metal-coated fullerenes, carbon nanotubes, carbon nanoparticles, and porous carbons. Since 1990 with the discovery of techniques to produce soluble carbon in a bottle (for examples, see Krätschmer et al. 1990 and references therein), research on and with carbon materials has skyrocketed (Dresselhaus et al. 1996; Dresselhaus and Dresselhaus 1995). Not only can the molecular forms of carbon (the fullerenes and their derivatives) be synthesized, characterized, and studied for applications, but many other new carbon materials such as multi- and single-walled carbon nanotubes can now be produced in macroscopic quantities. The broad variety of carbon nanotube structures whose properties are now being examined both theoretically and experimentally. A rich literature on these new carbon materials now exists. This report will only attempt to highlight a few important recent examples in the area of high surface area materials. Of particular interest for future catalytic applications is the recent report that not only can C60 be coated with metal atoms, but that the metal coating can consist of a precise number of metal atoms. For example, C60Li12 and C60Ca32 have been identified mass spectroscopically (Martin, Malinowski, et al. 1993; Martin, Naher, et al. 1993; Zimmerman et al. 1995). C60 has been coated with a variety of different metals, including Li, Ca, Sr, Ba, V, Ta and other transition metals. Interestingly, addition of more than 3 Ta atoms to C60 breaks the C60 cage. Replacement of one carbon atom in C60 by a transition metal atom such as Co or Ir is being studied for possible catalytic applications.

The future technological challenge will be to discover techniques to fabricate large quantities of such materials, so that such catalyst materials can be put in a bottle and not just in molecular beams. Examples of carbon nanotube structures, including multiwalled and metal-atomfilled nanotubes. Carbon nanotubes have the interesting property that they are predicted to be either semiconducting or conducting (metallic), depending on the chirality and diameter of the nanotube. Such materials are being studied as conductive additives to plastics and for use in electrochemical applications where the uniformity of the nanotube diameter and length is not overly critical (Dresselhaus 1998). Another approach is to use the carbon nanotube as a template for a nanotube of an inorganic oxide. Hollow nanotubes of zirconia and yttria-stablilized zirconia have been prepared by coating treated carbon nanotubes with a zirconium compound and then burning out the carbon template (Rao et al. 1997). Finally, large scale production of singlewalled nanotubes has recently been demonstrated, so one may anticipate a strong upsurge in the characterization and potential usage of single-walled carbon nanotubes in the future (Jounet et al. 1997). Porous carbons are of interest as molecular sieve materials, both as sorbants and as membranes, or as nanostraws for filtration. One of the major research objectives is to develop materials or structures with exceedingly high storage capacity per unit volume and weight for gases such as H2 or CH4. H2 or CH4 could become an economic source of combustion fuel or a means to power fuel cells for ultralow-emission vehicles or for electric power generation. Microporous hollow carbon fibers have exhibited high permeance and high selectivity as hydrogen selective membranes, and development is now underway to scale up these membranes to commercial levels (Soffer et al. 1987; Jones and Koros 1994; Rao and Sircar 1993). Carbon fiber materials produced via catalytic decomposition of hydrocarbon vapors have also recently been reported to exhibit exceptionally high hydrogen adsorption capacity (Baker 1998). More mundane uses of nanotubes are as nanometer reinforcing rods in polymers or even in concrete. Incorporation of conducting carbon nanotubes in construction materials such as concrete or structural plastics opens opportunities for real time monitoring of material integrity and quality.

Microporous and Dense Ultrathin Films

Research and development of microporous thin films for use as molecular sieving membranes using inorganic crystalline materials such as zeolites or porous silica is another area of active research around the world. For molecular sieving membranes, one critical challenge rests on discovering ways to create large scale, thin, nearly defect-free membranes. One recent example is the fabrication of mesoporous conducting thin films grown from liquid crystal mixtures (Attard et al. 1997). Transmission electron microscopy (TEM) reveals an ordered array of 2.5 nm diameter cylindrical holes in a 300 nm thick Pt film. The hole diameter can be varied either by changing the chain length of the surfactant molecule or by adding an alkane to the plating solution. It is interesting that this technique produces a continuous thin film with nanoscale porosity in an electrically conducting material. Dense ultrathin films such as single monolayer films would be of significant importance in the semiconductor industry. Thin films of specialized coatings for corrosion, thermal, and/or chemical stability should be valuable for the chemical and aerospace industries. Novel chemical sensors may be anticipated through use of ultrathin films composed of specialized clusters. Typical techniques for production of thin films are physical vapor deposition, chemical vapor deposition, and Langmuir- Blodgett processes.

Self-Assembly Processes for Nano Biotechnology

Researchers at the University of Massachusetts Amherst have developed a series of novel techniques in nanotechnology that hold promise for applications ranging from highly targeted pharmaceutical therapies, to development of nutrition-enhanced foods known as "nutraceuticals," to nanoscopic sensors that might one day advance medical imaging and diagnostics. The research, published in the Jan. 10 issue of Science, was funded by the U.S. Department of Energy and the National Science Foundation. The team included faculty members Thomas Russell and Todd Emrick of the department of polymer science and engineering, and Anthony Dinsmore of the department of physics, and graduate students Yao Lin and Habib Skaff, both of polymer science and engineering. "Our findings open new avenues to revolutionize technology by the controlled fabrication of nanoscopic materials having unique optical, magnetic and electronic properties," said Russell.

The study details three major findings:

- A novel method to create robust capsules from nanometer-sized particles
- A new technique to make nanoscopic particles water-soluble
- Functionalizing regions of the capsules with tailored properties, such as luminescence.

Emrick's research explores the behavior of nanoparticles to which ligands organic molecules and polymers have been attached. Russell is an expert in the surface and interfacial properties of polymers, and polymer-based nanostructures. Dinsmore specializes in colloidal assemblies and interface physics. "This is a productive collaboration in that we really have all the bases covered in terms of synthesis, understanding of interfacial activity and mediation, and the physics issues including surface tension and particle interactions," Emrick said. The study details a new method for assembling nanoparticles into robust, three-dimensional structures by encapsulating and stabilizing water droplets. Nanoparticles suspended in oil will self-assemble around a droplet of water, fully coating it with a shell. Although scientists have long known that particles tend to assemble at fluid interfaces, "the idea of using liquid interfaces as scaffolds is exciting and tremendously useful since researchers can tailor or modify the nanoparticles from both sides of the interface," explained Dinsmore. "We have much more surface area to work with for adding or removing specific particles." "Nanoparticles have exciting properties due to their small size, and they can be prepared in various shapes and sizes. What's really key is that you attach ligands that extend from the nanoparticles like hairs, in order to preserve the nanoscopic integrity of the particles and prevent them from clustering," Emrick said. "Changing the nature of these organic ligands can really modify the behavior of the particles. You can endow the nanoparticles, and thus the capsules that they form upon interfacial assembly, with a wide range of properties based on which ligands are attached." The effect of the ligands on the interactions of nanoparticles with the surrounding environment is crucial in medical applications. "These organic molecules will dictate the solubility, miscibility, and charge transport properties of the particles," Emrick said. UMass researchers also developed a method to take these nanoparticles, which are oil-soluble, and make them water-soluble, simply by shining light on them. "Developing nanoparticles that are water-soluble has significant implications for medicine in the biosensors area," Emrick said. "Using luminescent material, as we did, could lead to advances in very sophisticated medical-imaging techniques as the fluorescent nature of these particles allows them to be viewed and tracked over time." Finally, the UMass team discovered that when nanoparticles of different sizes compete for assembly at the interface, the bigger ones win, and segregate or cluster into patches on the droplet surface. "This opens a range of possibilities for developing nanoscopic capsules that have certain properties in specific areas," said Dinsmore. "You could build in an area with permeability, magnetism, or conductivity, so that one area would be functionally distinct."

In just a few decades physicians could be sending tiny machines into our bodies to diagnose and cure disease. These nanodevices will be able to repair tissues, clean blood vessels and airways, transform our physiological capabilities, and even potentially counteract the aging process. It is the year 2031, and the age of advanced nanomedicine has arrived. A young man arrives at his physician's office with a mild fever, nasal congestion, discomfort and a cough. The physician pulls from her pocket a lightweight, handheld device resembling a pocket calculator. She unsnaps from it a cordless, self-sterilizing, pencil-sized probe and inserts it into the patient's mouth as if it were a tongue depressor. On the tip of the probe are billions of molecular assay receptors, mounted on hundreds of self-guiding retractile stalks. Each receptor is sensitive to the chemical signature of a specific kind of bacterium or virus. "Ahhh," says the patient, and a few seconds later a three-dimensional, color-coded map of his throat appears on the display panel of the device.

Beneath the map scroll columns of data, revealing the unique molecular signature of a known and unwelcome bacterial pathogen. With the diagnosis complete, the infectious microorganism can be exterminated. No need for antihistamines, cough drops and a week-long course of antibiotics. The physician keeps several generic classes of nanorobots in her office for just such a circumstance.

Using a desktop appliance in her office, she programs billions of nanorobots to find, recognize and destroy the particular microbial strain. The nanomachines are suspended in a carrier fluid that the patient inhales into his lungs, after which the mobile devices march down the patient's throat, propelled on tiny legs. Following a search pattern, the nanorobots ingest and destroy the harmful bacteria they encounter using mechanical and chemical phagocytosis. The patient feels nothing: nanorobots are the size of bacteria, which constantly crawl on and inside the body without ever being noticed. After several minutes, the physician activates an acoustic homing beacon to guide the nanorobots back into the patient's mouth, where she retrieves them through a collection port on the tip of the homing device. A further survey with the original diagnostic probe reveals no evidence of the pathogen.

This is a wonderful vision of medicine in the future, but how do we get there from here? Nanotechnology ("nano" from the Greek word for dwarf) involves engineering and manufacturing at the molecular, or nanometer, level. One nanometer is one billionth of a meter, about the width of six bonded carbon atoms. How does one build at this scale? Scientists at New York University have adopted the self-assembly approach, and have succeeded in producing complementary strands of DNA that can zip themselves into complex structures. They have built cubes, octahedra and other solids made of just a few thousand nucleotides each, by the billions per batch.

Meanwhile at Cornell University, researchers have genetically engineered a natural biomotor normally found in the enzyme ATPase to allow it to incorporate nonbiological parts such as a silicon nitride bar 100 nanometers in length, making the first artificial hybrid nanomotor. In a microscopic video presentation, dozens of bars attached to artificial molecular motors in a large precise array could be seen spinning at 200 revolutions per minute, like a field of tiny propellers. Another way to build at the molecular level is by positional assembly—picking and placing molecular parts exactly where you want them. A device capable of positional assembly would work much like the robot arms that manufacture cars on automobile assembly lines, or like the ribosome that assembles amino acids one by one into proteins in our cells. The US engineering firm Zyvex Corp. is aiming to be the first to create an artificial molecular assembler using positional assembly to manufacture atomically precise structures.

Elsewhere, researchers at Harvard University have constructed the first general-purpose "nanotweezers," using a pair of electrically controlled carbon nanotubes. These have successfully grasped 300-nanometer clusters of polystyrene spheres and extracted a single semiconductor wire 20 nanometers wide from a mass of entangled wires. The scientists hope eventually to produce nanotweezers small enough to grab individual large molecules. Such nanoresearch will eventually lead to applications affecting almost every aspect of modern life, including the home, transport, computers, energy and the environment. Some of the most important and life-changing applications, however, are likely to be seen in medicine. We may see the first nanomedical materials and devices in use within the next few years. Scientists at the University of Michigan's Center for Biologic Nanotechnology are currently pursuing the use of dendrimers as a safer and more effective genetic therapy agent. Dendrimers are branching synthetic molecules that can be grown nanometer by nanometer to reach the desired size. They take on a spherical shape, and have sufficiently large openings and cavities to carry small molecules—dendrimers could therefore be used to sneak DNA into cells without triggering an immune response. At the same center, researchers have reported using dendrimer "nanodecoys" to trap and deactivate influenza virus particles. Relatively simple nanodevices could soon offer cures for major conditions such as diabetes. Researchers at Ohio State University and the University of Illinois have managed to construct silicon-based microcapsules highly perforated with "nanopores," each as small as 20 nanometers in diameter. These pores are large enough to allow small molecules such as oxygen, glucose and insulin to pass through, but are small enough to impede the passage of much larger immune system molecules such as immunoglobulins. Microcapsules containing islet cells could be implanted beneath the skin of some diabetes patients, temporarily restoring the body's delicate glucose control feedback loop without the need for powerful immuno-suppressants that can leave the patient at serious risk for infection.

The same method of encapsulation could be used to treat other enzyme- or hormone-deficiency diseases. Encapsulated neurons could also be implanted in the brain and then electrically stimulated to release neurotransmitters, possibly as part of a future treatment for Alzheimer's or Parkinson's disease. Artificial "biobots" could be in our bodies within five to ten years. Advances in genetic engineering are likely to allow us to construct an artificial microbe—a basic cellular chassis—to perform certain functions. These biobots could be designed to produce vitamins, hormones, enzymes or cytokines in which the host body was deficient, or they could be programed to selectively absorb and break down poisons and toxins. A new company called EngeneOS, Inc., founded in late 2000, has already announced plans to develop artificial Engineered Genomic Operating Systems using the techniques of molecular biology. These systems will comprise a library of component device modules and proprietary modular components. This will allow the engineering and construction of programmable biobots with novel form and function. The greatest power of nanomedicine will emerge in the longer term, perhaps 10-20 years from now, when we learn how to design and construct complete artificial nanorobots using strong diamond-like materials, nanometer-scale parts, and onboard subsystems including sensors, motors, manipulators, power plants, and molecular computers. One example is an artificial mechanical cell called a respirocyte, which could be used to keep a patient's tissues safely oxygenated for up to about four hours (at maximum dosage) if their heart has stopped beating. These cells (side panel, 4) could also enable a healthy person to sprint at top speed for at least 15 minutes without breathing, or to sit underwater at the bottom of a swimming pool for hours. Still entirely theoretical, the respirocyte is a micron-wide spherical nanorobot made of 18 billion atoms precisely arranged in a diamondoid structure to form a tiny 1,000-atmosphere pressure tank.

Several billion molecules of oxygen and carbon dioxide can be absorbed into, or released from, this tank using computer-controlled molecular pumps powered by serum glucose and oxygen. External gas concentration sensors would allow respirocytes to mimic the action of the natural hemoglobin-filled red blood cells, with oxygen released and carbon dioxide absorbed in the tissues, and vice versa in the lungs. Each respirocyte would be able to hold 200 times more gas per unit volume than a natural red cell, so a few cubic centimeters injected into the human bloodstream would exactly replace the gas carrying capacity of the patient's entire 5.4 liters of blood. Other proposed medical nanorobots offer equally astonishing performance improvements over nature. For instance, micronsize artificial mechanical platelets could allow complete hemostasis (control of bleeding by clot formation) in just one second, even for moderately large wounds, a response time 100-1,000 times faster than the natural system. This would be achieved by rapidly unfurling a compactly stowed onboard biodegradable mesh under the control of an onboard computer. These "clottocytes" would be about 10,000 times more effective as clotting agents than an equal volume of natural platelets. Nanorobotic phagocytes called microbivores could patrol the bloodstream, seeking out and digesting unwanted pathogens including bacteria, viruses or fungi. Each nanorobot could completely destroy one pathogen in just 30 seconds—about 100 times faster than natural leukocytes or macrophages—releasing a harmless effluent of amino acids, mononucleotides, fatty acids and sugars. No matter that a bacterium has acquired multiple drug resistance to antibiotics or to any other traditional treatment. The microbivore will eat it anyway, achieving complete clearance of even the most severe septicemic infections in minutes to hours, as compared to weeks or even months for antibiotic-assisted natural phagocytic defenses, without increasing the risk of sepsis or septic shock. Related nanorobots could be programed to recognize and digest cancer cells, or to clear circulatory obstructions within minutes in order to rescue stroke patients from ischemic damage.

More sophisticated medical nanorobots will be able to intervene at the cellular level, performing surgery within cells. Physician-controlled nanorobots could extract existing chromosomes from a diseased cell and insert newly manufactured ones in their place, a process called chromosome replacement therapy. This would allow a permanent cure of any pre-existing genetic disease, and permit cancerous cells to be reprogramed to a healthy state. In recent years, many gerontologists have begun to think of aging as a genetic disease that might be cured. Evidence from the Human Genome Project suggests that just a few hundred genes at most may be directly involved in aging. If we can completely understand these few genes and how they work, we may be able to alter them to eliminate this unwanted syndrome. Using cytosurgical nanorobots, corrected genes could be installed in every one of the 10 trillion tissue cells in our bodies. We would then no longer naturally age, and our bodies would again repair themselves as well as they did when we were children. Although nanotechnology is in its infancy, researchers are steadily making major breakthroughs. If we can learn to harness and precisely control the ability to manipulate molecules, then many aspects of our lives will change forever.

In particular, the ability to carry out medical procedures at the molecular level will revolutionize medical practice. The next few decades will be very interesting indeed. A new era on medicine are expected to happen in the coming years. Due to the advances in the field of nanotechnology, nanodevice manufacturing has been growing gradually. From such achievements in nanotechnology, and recent results in biotechnology and genetics, the first operating biological nanorobots are expected to appear in the coming five years, and more complex diamondoid based nanorobots will become available in about ten years. In terms of time it means a very near better future with significant improvements in medicine. In this work we present a practical approach taken on developing nanorobots for medicine in the sense of using computational nanomechatronics techniques as ancillary tools for investigating manufacturing design, nanosystems integration, sensing and actuation for medicine applications. Thus, the work describes pathways that could enable design testability, but also help scientists and profit corporations in providing the helpful information needed to test and design integrated devises and solutions towards manufacturing biomedical nanorobots. The use of robots in surgery has provided additional tools for surgeons enabling minimally invasive intervention or even long distance teleoperated surgeries. Indeed we may trust on human creativeness and technical capabilities that can ever be improved in terms of technical achievements. In recent years the medicine has enabled significant wellness for the life quality and longevity of the world population. And for the coming years, we may be prepared to experiment even more benefits, as results from advances that are being pursued step by step in new fields of science, such as nanobiotechnlogy. With the expected miniaturization of devices provided by several works on nanoelectromechanical systems (NEMS), nanomanufacturing has actually become a reality. Hence, with the NEMS recent advances on building nanodevices, and the development of interdisciplinary works, altogether may be translated in few years through the development of integrated nanomachines, also known as nanorobots. With the use of techniques that are advancing rapidly, such as nano-transducers, and biomolecular computing, nanorobots are expected to be able to operate in a well defined set of behaviors performing pre-programed tasks. Thus, in the coming few years, nanorobots being teleoperated to perform surgery, or even nanorobots continually supervising the human body in order to assist organs that may require some kind of repair, is one of the most expected revolutionary tools for biomedical engineering problems. The development of nanorobots is an emerging field with many aspects under investigation. Simulation is an essential tool for exploring alternatives in the organization, configuration, motion planning, and control of nanomachines exploring the human body.

Nanorobot applications could be focused mainly on two major areas, as follows: nanorobots for surgical interventions, as well as their utilization for patients that need constant monitoring. The nanorobots require specific controls, sensors and actuators, basically in accordance with each kind of biomedical problem. Advanced simulations can include various levels of detail, giving a trade-off between physical accuracy and the ability to control large numbers of nanorobots over relevant time scales with reasonable computational effort. Another advantage is that simulation can be done in advance of direct experimentation. It is most efficient to develop the control technology in tandem with the fabrication technologies, so that when we are able to build these devices, we will already have a good background in how to control them. Computational mechatronic approaches were proposed as a suitable way to enable the fast development of nanorobots operating in a fluid environment relevant for medical applications. Unlike the case of larger robots, the dominant forces in this environment arise from viscosity of low Reynolds number fluid flow and Brownian motion and such parameters are been implemented throughout a set of different investigations. Practical and innovative paradigms have been developed based on the Nanorobot Control Design (NCD) simulator that allows fast design testability comparing various control algorithms for nanorobots and their application for different tasks.

Also such information generated by the NCD can be useful as parameters for building nanodevices, such as transducers and actuators. In future decades the principal focus in medicine will shift from medical science to medical engineering, where the design of medically-active microscopic machines will be the consequent result of techniques provided from human molecular structural knowledge gained in the 20th and early 21st centuries. For the feasibility of such achievements in nanomedicine, two primary capabilities for fabrication must be fulfilled: fabrication and assembly of nanoscale parts. Through the use of different approaches such as biotechnology, supramolecular chemistry, and scanning probes, both capabilities had been demonstrated to a limited degree as early as 1998. Despite quantum effects which impart a relative uncertainty to electron positions, the quantum probability function of electrons in atoms tends to drop off exponentially with distance outside the atom. Even in most liquids at their boiling points, each molecule is free to move only ~0.07 nm from its average position. Developments in the field of biomolecular computing have demonstrated positively the feasibility of processing logic tasks by biocomputers, a promising first step toward building future nanoprocessors with increasing complexity. There has been progress in building biosensors and nanokinetic devices, which also may be required to enable nanorobotic operations and locomotion. Classical objections related to the feasibility of nanotechnology, such as quantum mechanics, thermal motions and friction, have been considered and resolved and discussions of techniques for manufacturing nanodevices are appearing in the literature with increasing frequency. One important challenge that has become evident as a vital problem in nanotechnology industrial applications is the automation of atomic-scale manipulation. The starting point of nanotechnology to achieve the main goal of building systems at the nanoscale is the development of control automation for molecular machine systems, which could enable the massively parallel manufacture of nanodevice building blocks. Governments all around the world are directing significant resources toward the fast development of nanotechnology. At least 30 countries have initiated activities in this field, and beyond that, with government investments to nanotechnology of US$ 800 million in Japan and US $ 774 million in USA. The U.S. National Science Foundation has launched a program in "Scientific Visualization" in part to harness supercomputers in picturing the nanoworld. A US $ 1 trillion market consisting of devices and systems with some kind of embedded nanotechnology is projected by 2015. More specifically, the firm 'display search' predicts rapid market growth from US $ 84 million today to $ 1.6 billion in 2007.

A first series of commercial nanoproducts has been announced as foreseeable by 2007. To reach the goal of building organic electronics, firms are forming collaborations and alliances that bring together new nanoproducts through the joint efforts of companies such as IBM, Motorola, Philips Electronics, Xerox/PARC, Hewlett-Packard, Dow Chemical, Bell Laboratories, and Intel Corp., among others. For such goals, new methodologies and theories to explore the nanoworld are the key technology. A useful starting point for achieving the main goal of building nanoscale devices is the development of generalized automation control for molecular machine systems which could enable a manufacturing schedule for positional nanoassembly manipulation. In our work we consider more specialized scheduling problems with a focus on nanomedicine: describing in a detailed fashion the nanorobot control designs and the surrounding virtual workspaces modelling that are required for the main kinematics aspects in the physically-based simulations.

Here the biomolecular assembly manipulation could be automatically performed by smart agents, which are given a set of possible tasks for biomedical engineering problems, embedded in a complex 3D environment. Virtual Reality could be considered as a suitable technique for nanorobot design and for the use of macro and microrobotics concepts given certain theoretical and practical aspects that focus on its domain of application. The collective nanorobotics approach proposed is one possible method to perform a massively-parallel positional nanoassembly manipulation . We constructed and demonstrated the applicability of multi-robot teams in timely sequenced set of works with practical applications that could enable the establishment of generalized control guidelines for nanorobotics. The use of multirobot teams working cooperatively to achieve a single global task applied to nanotechnology is a field of research that is very new and challenging. Research on collective robotics suggests that we should consider emulating the methods of the social insects to build decentralized and distributed systems that are capable of accomplishing tasks through the interaction of agents with the same structures and pre-programed actions and goals. Thus, a careful decomposition of the main problem task into subtasks with action based on local sensor-based perception could generate multirobot coherent behaviors. Several techniques was applied for such aims, as Neural Networks algorithms, Evolutionary computation, chemical based sensors and actuators, and even temperature time-gradients, just to quote a few.

Among other interesting nanorobot applications, we could foresee their use to process specific chemical reactions in the human body as ancillary devices for injured organs. Nanorobots equipped with nanosensors could be used to detect glucose demand in diabetes patients. Moreover, nanorobots could also be applied in chemotherapy to combat cancer through superior chemical dosage administration, and a similar approach could be taken to enable nanorobots to deliver anti-HIV drugs. Successful nanorobotic systems must be able to respond efficiently in real time to changing aspects of microenvironments not previously examined from a control perspective. Unlike some prior simulators for simple robots, in the present work we have developed nanorobots that are not restricted to a fixed grid nor behave as simple cellular automata with very simple environments. Also by contrast, most CAD approaches provide only animation or visualization tools, while the NCD is a physically-based simulator. The set of experiments we have carried out include the main physical properties existent in the environment where the nanorobots are being currently projected to be operating in the coming years. Hence, the majority of results from our investigation should be useful as well on analysing integrated manufacturing capabilities. In fact, including aspects of the physical environment in conjunction with graphical visualization provide a feasible approach for automation and control design. Furthermore, the architecture that was developed intended to enable the incorporation and evaluation of several control methods and distinct nanorobot shapes analyses. The automation, control, and manufacturing of nanorobots is a challenging and very new field. Realizing revolutionary applications of nanorobots to health or environmental problems raises new control challenges. The design and the development of complex nanomechatronic systems with high performance should be addressed via simulation to help pave the way for future applications of nanorobots in biomedical engineering problems.

Biomolecular Self-Assembly

Assembling nanoscale machinery and electronics poses a problem for which we have yet to find an adequate solution, hence nanotechnology has yet to arrive. However, that a solution to the problem exists can be seen in all living systems. Composed of molecular building blocks such as nucleic acids, proteins and phospholipids, biological organisms enter the world equipped with all the necessary means for assembling these components into extremely well organized structures. The purpose of this treatise is to summarize the basic tools that nature has provided for the nanoscientist. This vast supply of molecular components has been precisely designed not only to function as part of a complex molecular machine, but also with the ability to get themselves to the appropriate locations! Exactly how they find the right place is still under debate. While most agree that they bounce around until they stick somewhere, what exactly allows the speed and precision of most biomolecular self-assembly is still a mystery.

Biomimetic Self-Assembly

Langmuir-Blodgett (LB) films are formed when synthetic or naturally occurring surfactants are placed at the air-water (or liquid gas) interface where they can organize into a layer that is one molecule thick (a monolayer), two molecules thick (bilayer) or several molecules thick (multi-layer). The polar part of the surfactant molecule tends to be attached to the surface of the water while the nonpolar hydrophobic regions of neighboring molecules will stick together and orient themselves away from the surface. As the lateral pressure (2 dimensional pressure created by compressing the interface area) is changed, the film undergoes phase transitions corresponding to various levels of self-organization.

Self-Assembled Monolayers

Thiol self-assembly can be achieved by reacting thiol containing compounds with clean gold surfaces. Sulfhydryl groups on molecules will covalently bind to gold, thus allowing molecules to be arranged two dimensionally over a gold surface. This is extra useful because gold conducts electricity and makes for excellent electrical contacts, thus Scanning Tunneling Microscopy (STM) measurements can be made on such samples. For example, redox proteins can be mutagenically modified with cysteine residues (an amino acid that contains a thiol group). Exposing a gold surface to such engineered molecules results in self-assembled monolayers of protein.

Design and Control

Designing nanorobotic systems deal with vast variety of sciences, from quantum molecular dynamics, to kinematic analysis. The rules applicable to nanorobotics depend upon the nano material we intend to use. Nanomechanical robotic systems deal with science significantly different from the biological or inorganic nanorobotic systems. Eric Drexler, author of the famous book, *Nanosystems: Molecular machinery, manufacturing and Computation*, detailed on many areas of science which influences design of nanomechanical systems. In this review chapter on various laws which govern the designing of nanorobotics, we will concentrate on biological systems. We will consider that the components that details a nanorobot is made of biological components, such as, proteins and DNAs. There doesn't exist any particular guideline or a prescribed manner which details the methodology of designing a bio-nanorobot (bio-nanorobot implies nanorobots made up of biocomponents) up to the date. There are many complexities which are associated with using biocomponents (such as protein folding, presence of aqueous medium), but the advantages of using these are also quite considerable. These biocomponents offer immense variety and functionality at a scale where creating a manmade material with such capabilities would be extremely difficult. These bio components have been perfected by nature through millions of years of evolution and hence these are very accurate and efficient. As noted in the review section on Molecular Machines, F_1-ATPase is known to work at efficiencies which are close to 100 per cent. Such efficiencies, variety and form are not existent in any other form of material found today. Also, the other significant advantages in using protein-based bio-nano composites is the development and refinement over the last 30 years of tools and techniques that enable researchers to mutate proteins in almost any way imaginable. These mutations can consist of anything from simple amino acid side-chain swapping, to amino acid insertions or deletions, incorporation of non-natural amino acids, and even the combination of unrelated peptide domains into whole new structures. An excellent example of this approach is the engineering of the F_1-ATPase, which is able rotate a nanopropeller in the presence of ATP. A computational algorithm was used to determine the mutations necessary to engineer an allosteric zinc-binding site into the F_1-ATPase using site-directed mutagenesis. The mutant F_1-ATPase was then shown to rotate an actin filament in the presence of ATP with average torque of 34 pN nm. This rotation could be stopped with the addition of zinc, and restored with the addition of a chelator to remove the zinc from the allosteric binding site. This type of approach can be used for the improvement of other protein-based nanocomponents. Hence, these bio components seem to be a very logical choice for designing nanorobots. Some of the core applications of nanorobots are in the medical field and using biocomponents for these applications seems to be a good choice as they both offer efficiency and variety of functionality. This idea is clearly inspired by nature's construction of nanorobots, bacteria and viruses which could be termed as intelligent organisms capable of movement, sensing and organized control. Hence, our scope would be limited to the usage of these bio components in the construction of bio-nanorobotics. A roadmap is proposed which details the main steps towards the design and development of bio-nanorobots.

Step 1: Bio-nanocomponents

Development of bio-nanocomponents from biological systems is the first step towards the design and development of an advanced bio-nanorobot, which could be used for future applications. Since the planned systems and devices will be composed of these components, we must have a sound understanding of how these behave and how could they be controlled. From the simple elements such as structural links to more advanced concepts such as motors, each component must be carefully studied and possibly manipulated to understand the functional limits of each one of them. DNA and carbon nanotubes are being fabricated into various shapes, enabling possibilities of constructing newer and complex devices. These nanostructures are potential candidates for integrating and housing the bio-nanocomponents within them. Proteins such as rhodopsin and bacteriorhodopsin are a few examples of such bio-nanocomponents. Both these proteins are naturally found in biological systems as light sensors. They can essentially be used as solar collectors to gather abundant energy from the sun. This energy could either be harvested (in terms of proton motive force) for later use or could be consumed immediately by other components, such as the ATP Synthase nano rotary motor. The initial work is intended to be on the bio-sensors, such as, heat shock factor. These sensors will form an integral part of the proposed bio-nano assemblies, where these will be integrated within a nano structure and will get activated, as programed, for gathering the required information at the nano scale. Tools and techniques from molecular modeling and protein engineering will be used to design these modular components.

Step 2: Assembled Bio-nanorobots

The next step involves the assembly of functionally stable bio-nanocomposites into complex assemblies. Some examples of such complex assemblies or bio-nanorobots. A bio-nanorobot with its 'feet' made of helical peptides and its body of carbon nano tubes, while the power unit is a biomolecular motor. A conceptual representation of modular organization of a bio-nanorobot. The modular organization defines the hierarchy rules and spatial arrangements of various modules of the bio-nanorobots such as: the inner core (the brain/ energy source for the robot); the actuation unit; the sensory unit; and the signaling and information processing unit. By the beginning of this phase a "library of bio-nanocomponents" will be developed, which will include various categories such as, actuation, energy source, sensory, signaling etc. Thereon, one will be able to design and develop such bio-nanosystems that will have enhanced mobile characteristics, and will be able to transport themselves as well as other objects to desired locations at nano scale. Furthermore, some bio-nanorobots will have the capability of assembling various bio-components and nanostructures from in situ resources to house fabrication sites and storage areas, while others will just manipulate existing structures by repairing damaged walls or making other renovations. There will also be robots that not only perform physical labor, but also sense the environment and react accordingly. There will be systems that will sense an oxygen deprivation and stimulate other components to generate oxygen, creating an environment with stable homoeostasis.

Step 3: Distributive Intelligence, Programing and Control

With the individual bio-nanorobots in full function, they will now need to collaborate with one another to further develop systems and "colonies" of similar and diverse nanorobots. This design step will lay the foundation to the concept of bio-nano swarms (distributive bio-nanorobots). Here work has to be performed towards control and programing of bio-nanoswarms. This will evolve concepts like distributive intelligence in the context of bio-nanorobots. Designing swarms of bio-nanorobots capable of carrying out complex tasks and capable of computing and collaborating amongst the group will be the focus. Therefore, the basic computational architectures needs to be developed and rules need to be evolved for the bio-nanorobots to make intelligent decisions at the nano scale. To establish an interface with the macro world, the computers and electronic hardware have to be designed as well. The overall electronic communication architecture. From a location, humans should be able to control, monitor the behavior and action of these swarms. Also, the basic computational capabilities required for functioning of the swarms will be developed. A representative computational bio-nanocell, which will be deployed within a bio-nanorobot. This basic computational cell will initially be designed for data retrieval and storage at the nanoscale. This capability will enable us to program (within certain degrees of freedom) the swarm behavior in the bio-nanorobots. We will further be able to get their sensory data (from nano world) back to the macro world through these storage devices. This programming capability would form the core essence of a bio-nanorobotics system and hence enables them with immense power.

Step 4: Automatic Fabrication and Information Processing Machines

For carrying out complex missions, such as sensing, signaling and storing, colonies of these bio-nanorobotic swarms needs to be created. The next step in nanorobotic designing would see the emergence of automatic fabrication methodologies of such bio-nanorobots in vivo and in vitro. Capability of information processing, which will involve learning and decision making abilities, will be a key consideration of this step. This would enable bio-swarms to have capability of self-evolving based on the environment they will be subjected to. These swarms could be programed to search for alternate energy sources and would have an ability to adapt as per that resource. Energy management, self-repairing, and evolving will be some of the characteristics of these swarms.

Modular Organization: Modular organization defines the fundamental rule and hierarchy for constructing a bio-nanorobotic system. Such construction is performed through stable integration of the individual 'bio-modules or components', which constitute the bio-nanorobot. For example, if the entity ABCD, defines a bio-nanorobot having some functional specificity then, A, B, C, and D are said to be the basic bio-modules defining it. The basic construction will be based on the techniques of molecular modeling with emphasis on principles such as energy minimization on the hyper surfaces of the bio-modules; hybrid quantum-mechanical and molecular mechanical methods; empirical force field methods; and maximum entropy production in least time. Modular organization also enables the bio-nanorobots with capabilities such as, organizing into swarms, a feature, which is extremely desirable for various applications. The conceptual representation of modular organization. A more realistic scenario in which all the modules are defined in some particular spatial arrangements based on their functionality and structure. A particular module could consist of other group of modules, just like a fractal structure (defined as fractal modularity). The concept of bio-nanocode has been devised, which basically describes the unique functionality of a bio-nanocomponent in terms of alphabetic codes. Each bio-nanocode represents a particular module defining the structure of the bio-nanorobot. For instance, a code like E-M-S will describe a bio-nanorobot having capabilities of energy storage, mechanical actuation and signaling at the nanoscale. Such representations will help in general classifications and representative mathematics of bio nanorobots and their swarms. The proposed capabilities of the bio-modules along with their targeted general applications. The bio-nanocode EIWR || M || S || FG representing the bio-nanosystem.

The Universal Template—Bio-nano Stem System: The modular construction concept involves designing a universal template for bio-nano systems, which could be 'programed and grown' into any possible Bio-nano coded system. This concept mimics the embryonic stem cells found in the human beings, that are a kind of primitive human cells which give rise to all other specialized tissues found in a human foetus, and ultimately to all the three trillion cells in an adult human body. Our Bio-nano Stem system will act in a similar way. This universal growth template will be constituted of some basic Bio-nanocodes, which will define the Bio-Nano-stem system. This stem system will be designed in a manner that could enable it to be programed at run-time to any other required bio-module. One such variant of the Bio-nano stem system, having the Bio-nanocode: EIWR | | M | | S | | FG and having enhanced sensory abilities. Molecular modeling techniques in sync with extensive experimentations would form the basis for designing these bio-nano systems. As per the roadmap, various bio-nano components would have to be designed and then these functional elements would have to be bound into assemblies which could be controlled and programed. Some of the molecular modeling techniques we are emphasizing are described below.

Energy Minimization Methods: The potential energy depends upon the co-ordinates of the molecular configuration considered. In this method, points on the hypersurface (potential energy surface) are calculated for which the function has the minimum value. Such geometries which correspond to the minimum energy are the stable states of the molecular configuration considered. By this method we can also analyze the change is configuration of the system from one minimum state to another. Methods such as the Newton-Raphson and Quasi-Newton are employed to calculate these minima. This method of finding minimum energy points in the molecule is used to prepare for other advanced calculations such as molecular dynamics or Monte-Carlo simulations. It is further used to predict various properties of the system under study.

- *Molecular Dynamics Simulation Methods:* To begin predicting the dynamic performance (i.e. energy and force calculation) of a biocomponent (say a peptide) molecular dynamics (MD) is performed. This method utilizes Newton's Law of motion through the successive configuration of the system to determine its dynamics. Another variant of molecular dynamics employed in the industry is the Monte Carlo simulation, which utilizes stochastic approach to generate the required configuration of a system. Simulations are performed based on the calculation of the free energy that is released during the transition from one configuration state to the other (for example using the MD software CHARMM, Chemistry at Harvard Molecular Mechanics). In MD, the feasibility of a particular conformation of a biomolecule is dictated by the energy constraints. Hence, a transition from one given state to another must be energetically favorable, unless there is an external impetus that helps the molecule overcome the energy barrier. When a macromolecule changes conformation, the interactions of its individual atoms with each other—as well as with the solvent—constitute a very complex force system. With one of the aspects of CHARMM, it is possible to model say, a peptide based on its amino acid sequence and allow a transition between two known states of the protein using Targeted Molecular Dynamics (TMD). TMD is used for approximate modeling of processes spanning long time-scales and relatively large displacements. Say to achieve a large conformational change the biomolecule is 'forced' towards a final configuration 'F' from an initial configuration 'I' by applying constraints. The constraint is in the form of a bias in the force field. Molecular kinematic simulations are also being used to study the geometric properties and conformational space of the bio molecules (peptides). The kinematic analysis is based on the development of direct and inverse kinematic models and their use towards the workspace analysis of the bio molecules. This computational study calculates all geometrically feasible conformations of the bio molecule i.e. all conformations that can be achieved without any atom interference. This analysis suggests the geometric paths that could be followed by a bio molecule during the transition from the initial to a final state, while molecular dynamics narrow downs the possibilities by identifying the only energetically feasible paths. The workspace analysis also characterizes the geometrically feasible workspace in terms of dexterity.

Modular Pattern Recognition and Clustering Function: The instantaneous value of the property A of a molecular system can be written as, $A (p^n (t), r^n (t))$, where $(p^n (t)$ and $(r^n (t)$ represent the n momenta and positions of a molecular system at time t. This instantaneous value would vary with respect to the interactions between the particles. There are two parts to this function (termed as M function). One part (function A), evaluates and hence recognizes the equivalent modules (in term of properties) in the molecular system. The second part of the function (D) forms the cluster (like a bioisosteres, which are atom, molecules or functional groups with similar physical and chemical properties) of various modular patterns according to the characteristic behaviors as identified by function A and also tracks their variations with rest to time.

$$M^{o}\{A(p^n (t), r^n (t), C^n[n]), D (A (p,r,C,t), f (x, y, \ldots), Bn (t))\}$$

Where, A is the first part of the function; $(p^n (t)$ and $(r^n (t)$ represents the n momenta and positions of a molecular system at time t; and $C^n[n]$ is an n dimensional matrix element for the individual components of the molecular system, which would store the categorized values of the modular patterns. D is the second part of the function, and it takes on function A recursively. It also maps these clusters based on a fitness function and stores the time-variant value in sub-function B. This function is based on the hypothesis that in a complex molecular system, there are certain parts which have similar properties and behavior as the system goes from one state to the other. By identifying these property patterns we can considerably reduce the simulation and computational time frames and can have better predictability of the system. Having established the design for the bio-nano components, computational studies would focus on determining the overall design and architecture of the bio-nano assemblies. One of the methods used in the industry is of Molecular Docking.

Molecular Docking and Scoring Functions: Molecular docking method is very important for the design of nanorobotic systems. This method is utilized two fit two molecules together in 3D space. The method of molecular docking could be used to computationally design the bio-nanoassemblies. The algorithms like DOCK, genetic algorithms and distance geometry are being specifically explored. Many drug design companies are using depth-first systematic conformational search algorithms for docking, and therefore, is a primary candidate for current research. Docking algorithm generates a large number of solutions. Therefore, we require scoring functions to refine the results so produced. Scoring functions in use today approximate the "binding free energy" of the molecules. The free energy of binding can be written as an additive equation of various components reflecting the various contributions to binding. A typical equation would be:

$$DG_{bind} = DG_{solvent} + DG_{cont} + DG_{int} + DG_{rot} + DG_{t/r} + DG_{vib}$$

Where, $DG_{solvent}$ is the solvent effects contribution, cDG_{conf} is due to conformational changes in the biomolecule, DG_{int} due to specific molecular interactions, DG_{rot} due to restrictive internal rotation of the binding bio-molecules, $DG_{t/r}$ is due to loss in the translational and rotational free energies due to binding, and DG_{vib} is due to vibrational mode variations. Once the overall architecture to bind the bio-nanocomponents is carried out through molecular docking, we would need to combine the components into real assemblies. The molecular component binding is the next step in the design scheme.

Connecting bio-nano components in a binding site: There are typically two approaches to design connectors for binding the bio-nanocomponents. One method searches the molecular database and selects a connector based on geometrical parameters. CAVEAT is a traditional routine which employs this method. The other approach is to design a connector from the scratch atom by atom. The geometrical parameters being known (as the configuration and the architecture was defined by molecular docking) these connectors are designed. A typical method employs designing molecular templates (which do not interfere with the basic geometry and design of the binding bio-nanocomponents) for defining a basic connector for binding the bio-nanocomponents. Feasibility of designing these connectors in real life is one challenge faced by molecular modelers. Minimum energy criterion is employed via Metropolis Monte-Carlo simulation annealing method to select the basic structure of the final connector. Genetic algorithm approaches are also employed to generate such designs. The step 1 in the roadmap requires that we have a library of biocomponents which are characterized based on their potentials for using them for designing and developing nanorobots. Various molecular techniques could be employed for assessing the usefulness of the biocomponents. For developing a nanorobot, we require the knowledge of the structure of the biocomponents used and the degree of conformational change that each element is capable of undergoing, and the environmental cues (e.g., T, pH) that induce these structural changes. Based on the computational methods detailed in the previous section, the following experimental techniques could be used to obtain this information.

Circular dichroism spectroscopy (CD): CD spectroscopy can be utilized to gain insights into the secondary structure of each bio molecules such as peptides. In this technique, polarized far-UV light is used to probe the extent of secondary structure formation in solutions. This technique represents a proven method for estimating the structural composition of the nanorobotic elements during a change in environmental conditions. The effect of temperature on the structure of the nanorobotic elements can be evaluated by cooling the samples to +1°C and then heating to +85°C, with a five minute equilibration at each temperature interval. Intervals could be determined by first conducting CD scans with broad intervals, with subsequent narrowing as the location of the transition is identified. The effect of pH and ionic strength on the transitions could be monitored using separate, equilibrated preparations for each condition.

Förster Resonance Energy Transfer (FRET): Another method that can be used to rapidly assess the extent of conformational changes by the nanorobotic elements is FRET. In order to employ this method, the nanorobotic components have to be engineered to have cysteine residues at both ends. One end of the biocomponent then has to be labeled with a donor dye molecule (Alexa Fluor 488, Molecular Probes), and the other end labeled with an acceptor dye molecule (Alexa Fluor 594). The resulting FRET signal can be monitored with a laser (488 nm), whereby the signal changes would depend on the spatial separation of the ends of the biocomponent.

Nuclear Magnetic Resonance (NMR): NMR is utilized to determine the exact 3-D structure of the nanorobotic components. Although CD spectroscopy can be used to estimate the changes in the structural composition of the biocomponents, it does not provide the exact 3-D configuration. This information can be obtained through the use NMR spectroscopy and the labeling of the biocomponent with NMR-active atoms. In addition to determining the endpoint structures of the components, NMR can also be used to follow the kinetics of the biomolecule conformational changes, in real time.

Laser-Based Optical Tweezers (LBOT) and Single Molecule Fluorescence: LBOT and single molecule fluorescence could be used to estimate the forces exerted by the nanorobotic elements, and the structural changes that occur during the genesis of these forces, after exposure to different stimuli. LBOT can provide force and displacement results with a resolution of sub-picoNewton and approximately one nanometer, respectively. Single molecule fluorescence measurements, based on the previously described FRET method, can simultaneously provide information about the structural state of the molecule at each point of the force-displacement curve. To correlate force measurements with structural changes, combination of the LBOT method with FRET can be used. The recent development of this technique has enabled the simultaneous assessment of nanoscale structural changes and their associated biomechanical forces. For scanning and manipulating matter at the nanoscale, Scanning Tunneling Microscopy (STM), Scanning Electron Microscopy (SEM), Atomic Force Microscopy (AFM) and Scanned-Probe Microscope (SPM) seem to be the common tools. Current work is mainly focused on using atomic force microscopes (AFM) nanoprobes for teleoperated physical interactions and manipulation at the nanoscale. Precise manipulation could help scientists better understand the principles of nanorobots. To achieve this it is essential to visualize the atom-to-atom interaction in real-time and see the results in a fully immersive 3D environment. Also, to facilitate user input in nanorobotic systems it is essential to develop voice, gesture and touch recognition features in addition to the conventional visualization and manipulation techniques. Virtual Reality (VR) technology is applied here, which not only provides immersive visualization but also gives an added functionality of navigation and interactive manipulation of molecular graphical objects. VR technology comes to our aid by providing the experience of perception and interaction with the nano world through the use of sensors, effectors and interfaces in a simulated environment. These interfaces transform the signals occurring at nanoscale processes into signals at macro-level and vice-versa. The requirement is that the communication with the nano-world must be at high-level and in real-time, preferably in a natural, possibly intuitive 'language'. Considering the nano-specific problems related to task application, tools and the interconnection technologies it leads to many flexible nano manipulation concepts. They can range from pure master/slave tele-operation (through 3-D visual/VR and/or haptic force feedback), over shared autonomy control (where for example some degrees of freedom are tele-operated and other are operating autonomously) to fully autonomous operation. In order to precisely control and manipulate biomolecules, we need tools that can interact with these objects at the nanoscale in their native environments.

Existing bio-nanomanipulation techniques can be classified as non-contact manipulations including laser trapping and electro-rotation, and contact manipulation referred to as mechanical stylus-, AFM- or STM-based nanomanipulation. The rapid expansion of AFM studies in biology/biotechnology results from the fact that AFM techniques offer several unique advantages: first, they require little sample preparation, with native biomolecules usually being imaged directly; second, they can provide a 3-D reconstruction of the sample surface in real space at ultra-high resolution; third, they are less destructive than other techniques (e.g. electron microscopy) commonly employed in biology; and fourth, they can operate in several environments, including air, liquid, and vacuum. Rather than drying the sample, one can image quite successfully with AFM in fluid. The operation of AFM in aqueous solution offers an unprecedented opportunity for imaging biological molecules and cells in their physiological environments and for studying biologically important dynamic processes in real time. Currently, these bio-nanomanipulations are conducted manually; however, long training, disappointingly low success rates from poor reproducibility in manual operations, and contamination call for the elimination of direct human involvement. Furthermore, there are many sources of spatial uncertainty in AFM manipulation, e.g., tip effects, thermal drift, slow creeping motion, and hysteresis. To improve the bio-nanomanipulation techniques, automatic manipulation must be addressed. Visual tracking of patterns from multiple views is a promising approach, which is currently investigated in autonomous embryo pronuclei DNA injection. Interactive nanomanipulation can be improved by imaging 3D viewpoint in a virtual environment. Construction of a VR space in an off-line operation mode for trajectory planning combined with a real time operation mode for vision tracking of environmental change ensures a complete 'immersed' visual display. An example of 3D bio-micro/nano manipulation system with a 3D VR model of the environment including the biocell, and carrying out the user viewpoint change in the virtual space.

Molecular Dynamics Simulations in Virtual Environment: Molecular Dynamics (MD) simulations of complex molecular systems require enormous computational power and produces large amount of data in each step. The resulting data includes number of atoms per unit volume, atomic positions; velocity of each atom, force applied on each atom and the energy contents. These results of MD simulations need to be visualized to give the user a more intuitive feel of what is happening. Haptic interaction used in conjunction with VR visualization helps the scientist to control/monitor the simulation progress and to get feedback from the simulation process as well.

The control of nanorobotic systems could be classified in two categories:

(i) Internal control mechanisms.
(ii) External control mechanisms.

The other category could be the hybrid of internal and external control mechanisms. This type of control depends upon the mechanism of biochemical sensing and selective binding of various biomolecules with various other elements. This is a traditional method, which has been in use since quite sometime for designing biomolecules. Using the properties of the various biomolecules and combining with the knowledge of the target molecule that is to be influenced, these mechanisms could be effective. But again, this is a passive control mechanism where at run time these biomolecules cannot change their behavior. Once programed for a particular kind of molecular interaction, these molecules stick to that. Here lies the basic issue in controlling the nanorobots which are supposed to be intelligent and hence programed and controled so that they could be effective in the ever dynamic environment. The question of actively controling the nanorobots using internal control mechanism is a difficult one. We require an 'active' control mechanism for the designed nanorobots such that they can vary their behavior based on situations they are subjected to, similar to the way macro robots perform. For achieving this internal control, the concept of molecular computers could be utilized. Leonard Adleman (from the University of Southern California) introduced DNA computers a decade ago to solve a mathematical problem by utilizing DNA molecules. Professor Ehud Shapiro's lab (at Israel's Weizmann Institute) has devised a biomolecular computer which could be an excellent method for an internal 'active' control mechanism for nanorobots. They have recently been successful in programing the biomolecular computer to analyze the biological information, which could detect and treat cancer (prostate and a form of lung cancer) in their laboratory. The molecular computer has an input and output module which acting together can diagnose a particular disease and in response produce a drug to cure that disease. It uses novel concept of software (made up of DNAs) and hardware (made up of enzymes) molecular elements.

This molecular computer is in generalized form and can be used for any disease which produces a particular pattern of gene expression related to it. This type of control mechanism employs affecting the dynamics of the nanorobot in its work environment through the application of external potential fields. Researchers are actively looking at using MRI as an external control mechanism for guiding the nano particles. Professor Sylvain Martel's Nanorobotics Laboratory (at École Polytechnique de Montréal, Canada) is actively looking at using MRI system as a mean of propulsion for nanorobots. An MRI system is capable of generating variable magnetic field gradients which can exert force on the nanorobot in the three dimensions and hence control its movement and orientation. Professor Martel's laboratory is exploring effect of such variable magnetic field on a ferromagnetic core that could probably be embedded into the nanorobots. Other possibilities being explored are in the category of 'hybrid' control mechanisms where the target is located and fixed by an external navigational system but the behavior of the nanorobot is determined locally through an active internal control mechanism. The use of nanosensors and evolutionary agents to determine the nanorobots behavior is suggested by the mentioned reference. Manipulating matter at molecular scale and influencing their behavior (dynamics and properties) is the biggest challenges for the nanorobotic systems. This field is still in very early stages of development and still a lot has to be figured out before any substantial outcome is produced. The recent explosion of research in nanotechnology, combined with important discoveries in molecular biology have created a new interest in bio nanorobotic systems.

The preliminary goal in this field is to use various biological elements—whose function at the cellular level creates a motion, force or a signal—as nanorobotic components that perform the same function in response to the same biological stimuli but in an artificial setting. In this way proteins and DNA could act as motors, mechanical joints, transmission elements, or sensors. If all these different components were assembled together they can form nanorobots and nanodevices with multiple degrees of freedom, with ability to apply forces and manipulate objects in the nanoscale world, transfer information from the nano to the macroscale world and even travel in a nanoscale environment. Ability to determine the structure, behavior and properties of the nanocomponents is the first step which requires focused research thrust. Only when the preliminary results on these nanocomponents are achieved, steps towards actually building complex assemblies could be thought of. Still problems like protein (basic bio nano component folding, precise mechanism behind working of the molecular motors like ATP Synthase have still has to be solved. Active control of nanorobots has to be further refined. Hybrid control mechanisms, where in, a molecular computer and external (navigational) control system work in sync to produce the precise results seems very promising. Further, concepts like swarm behavior in context of nanorobotics is still have to be worked out. As it would require colonies of such nanorobots for accomplishing a particular task, concept of co-operative behavior, distributed intelligence has to be evolved. The future of bio-nanorobots (molecular robots) is bright. We are at the dawn of a new era in which many disciplines will merge including robotics, mechanical, chemical and biomedical engineering, chemistry, biology, physics and mathematics so that fully functional systems will be developed. However, challenges towards such a goal abound. Developing a complete database of different biomolecular machine components and the ability to interface or assemble different machine components are some of the challenges to be faced in the near future.

However broad the range of synthesis approaches, the critical control points fall into two categories:

1. control of the size and composition of the nanocluster components, whether they are aerosol particles, powders, semiconductor quantum dots, or other nanocomponents
2. control of the interfaces and distributions of the nanocomponents within the fully formed materials These two aspects of nanostructure formation are inextricably linked; nevertheless, it is important to understand how to exercise separate control over the nucleation of the nanostructure building blocks and the growth (for example, minimizing coagulation or agglomeration) of those components throughout

the synthesis and assembly process. This latter issue is related to the importance of the following:

- the chemical, thermal, and temporal stability of such formed nanostructures
- the ability to scale-up synthesis and assembly strategies for low-cost, large-scale production of nanostructured materials, while at the same time maintaining control of critical feature size and quality of interfaces (economic viability is a compelling issue for any nanostructure technology) All researchers in this area are addressing these issues.

Outstanding Issues: Opportunities and Challenges

The broad explorations of this WTEC study have shown a great diversity of impressive work on nanostructured materials. One could argue that many aspects of the work on nanostructured materials have been long-established efforts, with well-developed techniques that have been brought forth into the manufacturing arena. How do we then explain the current excitement and interest in nanostructured materials as a promising new endeavor? Part of the answer lies in the recognition of common scientific issues and common enabling technologies that link this group of researchers together. Ready availability of ever more sophisticated characterization methods that allow us to visualize and probe materials at the nanoscale has accelerated the pace of activities in this field; at the same time, recognition of the common critical issues of control over nanostructure size and placement motivates sharing of solutions over the boundaries of conventional disciplines.

Thus, some of the most exciting findings of this study manifest the crossfertilization of techniques and ideas; for example, aerosol particles are being integrated with more traditionally fabricated electronic nanostructures, with placement achieved through the manipulation of STM tips, in order to explore ideas of electronic device enhancement at the nanostructure level (Junno et al. 1998). The ordered assembly of diblock copolymers define three-dimensional nanostructures, and those structures are transferred into electronic (semiconductor) substrates through high resolution pattern transfer processes (Möller 1998). Recognizing the need for effective utilization of ideas across disciplines brings the responsibility of establishing the educational infrastructure that will adequately train young scientists to more fully develop the concepts and applications of nanostructured materials in the future.

Such an infrastructure will require researchers and educators who are familiar with the properties of a broad range of materials, including polymers, biomolecular materials, metals, ceramics, and semiconductors. It will benefit from an informed perspective on critical applications, and must provide access to a wide range of synthesis and assembly techniques and characterization methods that currently reside separately in the disciplines of physics, chemistry, electrical engineering, biology, etc.

Finally, a critical enabler for the future of this field and for its educational infrastructure is further development of computational tools that encompass the full range of atomistic calculations to macroscopic materials properties. This will require a "systems approach" (Olson 1997) that will span a variety of different computational models, addressing the different critical length scales, starting with solutions of the Schrodinger and Poisson equations, solving interatomic forces, and scaling all the way up to simulations of macroscopic properties and behavior (Goddard 1998). Increased appreciation of and access to the diverse means of nanoparticle synthesis and assembly that have been developed within many different disciplines, and a common development of enabling tools and technologies, will enhance the pace of accomplishments in this new area of nanoscale synthesis and assembly.

Bibliography

Avery, O.T., MacLeod, C.M.& McCarty, M. Studies of the chemical nature of the substance inducing transformation of pneumococcal types. Induction of transformation by a desoxyribonucleic acid fraction isolated from *Pneumococcus* Type III. *J. Exp.Med.* 79, 137–158 (1944).

Bush, G.W., "Stem Cell Science and the Preservation of Life, " *The New York Times*, August 12, 2001, p. D13.

Collins, F. & Galas, D. A new five-year plan for the US Human Genome Project. *Science* 262, 43–46 (1993).

Collins, F.S. *et al.*New goals for the US Human Genome Project: 1998–2003. *Science* 282, 682–689 (1998).

Drews, J.& Ryser, S. The role of innovation in drug development.

Fuller, B.P. *et al.* Policy forum: Ethics — privacy in genetics research. *Science* 285, 1359–1361 (1999).

Gabriel, S. B. *et al.* The structure of haplotype blocks in the human genome. *Science* 296, 2225–2229 (2002).

Gardner, M.J. *et al.* Genome sequence of the human malaria parasite *Plasmodium falciparum*. *Nature* 419, 498–511 (2002).

Gavin, A.C. *et al*. Functional organization of the yeast proteome by systematic analysis of protein complexes. *Nature* 415, 141–147 (2002).

Golub, T.R.Genomic approaches to the pathogenesis of hematologic malignancy.*Curr.Opin.Hematol.* 8, 252–261 (2001).

Guttmacher and Mark S.Guyer are at the National Human Genome Research Institute, National Institutes of Health, Bethesda, Maryland 20892, USA.

Guttmacher, A.E. & Collins, F.S. Genomic medicine — A primer.*N.Engl. J.Med.* 347, 1512–1520 (2002).

Hannon, G.J. RNA interference. *Nature* 418, 244–251 (2002).

Hilbert, D. Mathematical problems. *Bull.Am.Math. Soc.* 8, 437–479 (1902).

Hirschhorn, J.N., Lohmueller, K., Byrne, E. & Hirschhorn, K.A comprehensive review of genetic association studies. *Genet.*

Holt, R.A. *et al.* The genome sequence of the malaria mosquito *Anopheles gambiae*. *Science* 298, 129–149 (2002).

Horii, T., Nagao, Y., Tokunaga, T., and Imai, H. 20003. Serum-free culture of murine primordial germ cells and embryonic germ cells. Theriogen. 59, 1257-1264.

Jackson, D. A., Symons, R. H. & Berg, P. Biochemical method for inserting new genetic information into DNA of Simian Virus 40: circular SV40 DNA molecules containing lambda phage genes and the galactose operon of *Escherichia coli*. *Proc.Natl Acad. Sci.*

Kaczorowski, D.J., Patterson, E.S., Jastromb, W.E., Shamblott, M.J. 2002. Review Article. Glucose-responsive insulin-producing cells from stem cells. Diabetes Metab.

Kalka, C. et al. Transplantation of ex vivo expanded endothelial progenitor cells for therapeutic neovascularization. Proc. Natl. Acad. Sci. USA 2000; 97:3422-3427.

Kalkman EAJ, Bilgin YM, van Haren P, van Suylen R-J, Saxena PR, Schoemaker RG.Determinants of coronary reserve in rats subjected to coronary artery ligation or aortic banding. Cardiovasc Res 1996.

Kocher AA, Schuster MD, Szabolcs MJ, Takuma S, Burkhoff D, Wang J, Homma S, Edwards NM, Itescu S. Neovascularization of ischemic myocardium by human bone-marrow-derived angioblasts prevents cardiomyocyte apoptosis, reduces remodeling and improves cardiac function. Nature Med.

L Field: Future therapy for cardiovascular disease. In Proceedings of the NHLBI Workshop Cell Transplantation: Future Therapy for Cardiovascular Disease? Columbia, MD; 1998.

Labastie M-C, Cortes F, Romeo P-H, Dulac C, Peault B. Molecular identity of hematopoietic precursor cells emerging in the human embryo. Blood 1998;92:3624-3635.

Mendel, G.Versuche über Pflanzen-Hybriden.*Verhandlungen des naturforschenden Vereines, Abhandlungen, Brünn* 4, 3–47 (1866).

Milki, A.A. et al., "Incidence of monozygotic twinning with blastocyst transfer compared to cleavage-stage transfer" *Fertility and Sterility* 79(3): 503-506 (2003).

Norwitz, E. R., et al., "Implantation and the survival of early pregnancy, " *New England Journal of Medicine* 345(19): 1400-1408 (2001)

Ogawa M, Kizumoto M, Nishikawa S, Fujimoto T, Kodama H, Nishikawa SI. Blood 1999;93:1168-1177.

Orlic D, Kajstura J, Chimenti S, et al. Bone marrow cells regenerate infarcted myocardium. Nature 2001, 410:701-705.

Oz MC, Argenziano M, Catanese KA, et al. Bridge experience with long-term implantable left ventricular assist devices. Are they an alternative to transplantation? Circulation. 1997; 95:1844-1852.

Pouzet B, Ghostine S, Vilquin J-T, Garcin I, Scorsin M, Hagege AA, Duboc D, Schwartz K, Menasche P (2001) Is skeletal myoblast transplantation clinically relevant in the era of angiotensin-converting enzyme inhibitors? Circulation 104:I223-228.

Rafii S, Shapiro F, Rimarachin J, Nachman R, Ferris B, Weksler B, Moore AS, Asch AS. Isolation and characterization of human bone marrow microvascular endothelial cells: hematopoietic progenitor cell adhesion.

Smythe GM, Grounds MD (2000) Exposure to tissue culture conditions can adversely affect myoblast behaviour in vivo in whole muscle grafts: implications for myoblast transfer therapy. Cell Transplant 9: 379-393.

Smythe G.M., Hodgetts S.I., Grounds M.D. (2001) Problems and solutions in myoblast transfer therapy. J Cell Mol Med 5:33-47.

Stockwell, B. R. Chemical genetics: Ligand-based discovery of gene function. *Nature Rev. Genet.* 1, 116–125 (2000).

Index

E

F

G

H

I

J

K